Y0-ARV-142

Jerry Gaston

Editor

Sociology of Science

Jossey-Bass Publishers
San Francisco • Washington • London • 1978

SOCIOLOGY OF SCIENCE
Problems, Approaches, and Research
by Jerry Gaston, Editor

Jossey-Bass, Inc., Publishers
433 California Street
San Francisco, California 94104

Jossey-Bass Limited
28 Banner Street
London EC1Y 8QE

Sociology of Science, edited by Jerry Gaston, is a special double issue of *Sociological Inquiry* (Volume 48, Nos. 3–4, 1978), journal of the National Sociology Honor Society, a quarterly publication of the United Chapters of Alpha Kappa Delta.

Library of Congress Catalogue Card Number LC 78-70090

International Standard Book Number ISBN 0-87589-397-X

Manufactured in the United States of America

JACKET DESIGN BY WILLI BAUM

FIRST EDITION

Code 7834

The Jossey-Bass
Social and
Behavioral Science Series

Preface

Since the early 1960s the sociology of science has become a recognized research specialty that has received increasing interest and has drawn, in relative terms, a large percentage of new recruits into its ranks. As one of these recruits, I was not reticent about accepting an invitation from Harry Johnson, editor of *Sociological Inquiry,* to edit a special double issue focusing on the sociology of science. We decided to include three kinds of chapters, which are presented in each of the three parts: the normative order in science, problem choice in science, and different approaches in the sociology of science. We had two objectives: (1) to provide a selection that would interest scholars in areas different from the sociology of science, showing the type of theories and problems that specialists currently are trying to explore; (2) to provide colleagues in the sociology of science with important, original work. Although some chapters began as papers presented at scientific meetings, they appear in final form here for the first time. I hope, and expect, that these papers will stimulate discussion, further study, and perhaps more than a little controversy.

This volume is designed for all scholars with interests in

the social aspects of science: historians, philosophers, economists, political scientists, sociologists, as well as physical and biological scientists.

Appreciation is due several people who made the publication of these chapters possible, and on a timely schedule. I thank Harry Johnson for his assistance at every stage of the endeavor. Bernard Barber assisted to a greater degree than he, with characteristic modesty, would want to admit. (He is not, of course, responsible for the inevitable shortcomings.) The contributors cooperated in a manner that is unusual in scholarly circles; the last chapter arrived well before the deadline for sending the manuscripts to the publisher. And Beverly Morber, my excellent secretary and assistant, performed in her usual extraordinary manner, which, even after some years, continues to amaze me. I acknowledge her assistance with much gratitude.

Carbondale, Illinois
September 1978

JERRY GASTON

Contents

Contributors

JERRY GASTON is professor of sociology and chairman of the department of sociology at Southern Illinois University (Carbondale), where he has taught since 1969. He attended East Texas State University, receiving a B.A. degree in social science (1962) and an M.A. degree in government (1963). During the 1963–1965 academic years, he was instructor in sociology at ETSU. From 1965 he attended Yale University, receiving the M.Phil. (1967) and Ph.D. (1969) degrees in sociology.

He is the author of *Originality and Competition in Science* (1973); *The Sociology of Science in Europe,* edited with Robert K. Merton (1977); and *The Reward System in British and American Science* (1978). In July 1978, he began a three-year term as editor of *The Sociological Quarterly.*

DONALD DEB. BEAVER has a Ph.D. degree in the history of science from Yale University and has published many papers in the field. Currently he is professor and chairman of the history of science at Williams College.

JOSEPH BEN-DAVID is professor of sociology at the Hebrew University in Jerusalem. His many publications include *Fundamental Research and the University* (1968), *The Scientist's Role in Society* (1971), *Trends in American Higher Education* (1974), and *Centers of Learning: Britain, France, Germany, United States* (1977).

THOMAS F. GIERYN, having completed graduate study at Columbia University, has accepted a position in the department of sociology at Indiana University. His chapter is based on part of his doctoral dissertation, "Problem Choice in Science: American Astronomers, 1951–75."

LOWELL HARGENS, who earned his doctorate at the University of Wisconsin, is associate professor of sociology at Indiana University. Among his publications is a book in the Rose Monograph Series, *Patterns of Scientific Research* (1975).

WILLIE PEARSON, JR., a doctoral candidate at Southern Illinois University, is working on a thesis involving black scientists in the United States.

DONALD R. PLOCH, professor and head of the department of sociology at the University of Tennessee, was director of the sociology program at the National Science Foundation from 1972 to 1977.

DEREK DE SOLLA PRICE, Avalon Professor of the History of Science at Yale University, has authored numerous books and papers. His latest book, edited with Ina Spiegel-Rösing, is *Science, Technology and Society: A Cross-Disciplinary Perspective* (1977).

BARBARA F. RESKIN, with a Ph.D. degree from the University of Washington, is associate professor at Indiana University. She has published papers in the *American Sociological Review, American Journal of Sociology,* and *Social Forces.*

NICO STEHR is associate professor of sociology at the University of Alberta (Canada). His recent book, edited with René König, is *Wissenschaftssoziologie* [Sociology of Science] (1975).

HARRIET ZUCKERMAN is professor of sociology, and she chairs the department of sociology at Columbia University. She has published many papers and is the author of *Scientific Elite: Nobel Laureates in the United States* (1977) and co-editor of *Toward a Metric of Science* (1978).

Sociology of Science

Problems, Approaches, and Research

Part 1

The Norm of Universalism

The sociology of science is well established now as a specialty in research and in teaching. But it was not always this way. In the foreword to Bernard Barber's (1952) *Science and the Social Order,* Robert Merton wrote, "This field has long remained in a condition of remarkable neglect" (p. 7). However, he noted, "When something is widely defined as a social problem in modern Western society, it becomes a proper object for study" (p. 17). And science is now considered by many to be, if not a social problem itself, the cause of significant social problems. Concern about science from a sociological perspective has produced the development of a specialty that, characteristically, has followed paths similar to the development of scientific specialties generally (see Cole and Zuckerman, 1975).

Thus, in due course, the sociology of science arrived. And in 1975, in cooperation with other scholars interested in the social aspects of science, sociologists helped organize the Society for the Social Studies of Science (4S). With considerable international interest, membership has grown to about five hundred, and its third annual meeting was held in November 1978.

The development of the sociology of science has been described from various perspectives (see Cole and Zuckerman, 1975; Merton, 1977). Several books published recently have highlighted the growing topics that sociologists of science are investigating (see Whitley, 1974; Stehr and König, 1975; Knorr and others, 1975; Blume, 1977; Mendelsohn and others, 1977; Merton and Gaston, 1977; Jones, 1978). As the sociology of science has developed, it has experienced considerable intellectual controversy and debate.

At the beginning, the primary theoretical perspective involved what has come to be called Mertonian sociology of science. As the sociologist responsible for maintaining a long-term interest in the subject, and as one who has offered significant ideas that led to early investigation, Merton is often considered the unofficial "founder" of the field.

Mertonian sociology of science is based on the idea that the social institution of science can be studied as any other social institution. As an institutionalized activity, science has goals and proceeds, through the various social processes, to realize its goals. This conception naturally leads to a variety of questions familiar to all sociologists.

Merton's norm of universalism provides the question in science that sociologists generally have found interesting: What are the differences between a theoretically ideal world and the empirically real world? If universalism in science precludes consideration of ascriptive characteristics, then these irrelevant characteristics should not help explain anything about the processes and outcomes in science. Race and sex are two ascribed statuses that are particularly interesting because of the norm of universalism.

For some years, and no doubt encouraged by the women's movement, studies of many kinds have asked the question: Are women discriminated against in the academic world (and in science) and, if so, to what extent? The question that should be added, but has not been asked systematically, is: "With what *effect* for science?" Studies on sex status and science have not reached similar conclusions. Some studies suggest that, on the basis of certain data, women suffer severe discrimination; others conclude that the effect of sex status is ambiguous. Some of these studies—it simply must be said—were executed in a manner that does not equal the desirable level of sociological competence, and sociologists should be pleased that some of these studies were not executed by sociologists.

A familiar strategy has been to consider marginal distributions and conclude that sex discrimination exists. For example, if data on rank distribution by sex in universities show that women hold a smaller percentage of professorships than men, this finding does not prove that women are discriminated against. Indeed, it could suggest correctly that women are given preferred treatment, because more women have been hired in recent years and have not had time yet to be promoted to higher ranks. The point of this illustration is neither to argue with the good evidence that strongly suggests discrimination nor to deny that discrimination exists. The purpose is to mention the necessary caution that must be used when one is confronted with data that do not control for the variables relevant to a social process. In organizations where seniority is the primary criterion for promotion to higher positions, only a few variables may have to be considered, but that is not the case in universities with research missions. Role performance (and its evaluation) is important and can be assessed for members in the organization. At the end of a study an author may, from inadequate data, correctly conclude that discrimination actually exists. But these studies must be excluded from important discussions except as object lessons for improving inquiry into the really important question of discrimination.

Fortunately, Barbara F. Reskin's chapter is an exemplar of the proper strategy in looking at the issue of sex status in science. Her purpose here is not to review the state of the art of the sex-role literature. Rather, she focuses on the types of questions that should be raised in a comprehensive study of the effect of sex status on a researcher's role performance in the total context of a scientific career. Her chapter is a comprehensive theoretical statement on the important issue of sex status in science.

The scientific community in the United States is almost unique in having majority and minority representation, and especially so when the minority in question has been represented in the population for over three hundred years. But, because there are relatively few black scientists (about 1 percent of scientists are black

Americans), there has been little study of their status. Willie Pearson, Jr., reviews this situation and describes the kinds of data that are necessary to deal with the question of universalism in science—when one is concerned with the neglected variable of racial identity. Moreover, he suggests the types of problems that black scientists, if they existed in sufficient number, might investigate. Thus, Pearson's paper involves both the theoretical problem of universalism in science, which asks about the conditions that might produce particularism, and the problem of the connection between social position and choice of research topics.

Questions about the norms of universalism are also the focus of Donald R. Ploch's chapter. Recently in the United States, the system of peer review has been questioned critically. The Congress has informants who have raised questions, with implicit and sometimes explicit opinions, about the fairness of federal funding. The distribution of funds is skewed, as shown by marginal distributions—the same level of inquiry some people are fond of using to "prove" discrimination against women in universities.

Of course, the policy for distribution of federal funds for research can be changed by the Congress. Instead of facing the question squarely, however, and dealing with the variables relevant to various specifications of the questions, critics charge the peer review system with favoritism, cronyism, and other qualities equally undesirable in a democratic society. The peer review system has produced a distribution of funds that displeases many people; but—instead of suggesting that government policy should be formulated to distribute research funds on some formula basis, without regard to competence, creativity, potential, and so forth—critics have attached great importance to perceived inequities or, in extreme cases, to a few valid examples of peer review abuse which are the inevitable slippage in any social system. Ploch's chapter demonstrates how theoretical ideas in the sociology of science, combined with appropriate methodology, can answer questions that have intellectual significance and policy implications.

References

Barber, B. *Science and the Social Order.* New York: Free Press, 1952.

Blume, S. *Perspectives on the Sociology of Science.* London: Wiley, 1977.

Cole, J., and Zuckerman, H. "The Emergence of a Scientific Specialty: The Self-Exemplifying Case of the Sociology of Science." In L. A. Coser (Ed.), *The Idea of Social Structure*. New York: Harcourt Brace Jovanovich, 1975.

Jones, R. A. (Ed.). *Research in the Sociology of Knowledge, Sciences, and Art*. Greenwich, Conn.: JAI Press, 1978.

Knorr, K., and others (Eds.). *Determinants and Controls of Scientific Development*. Boston: Reidel, 1975.

Mendelsohn, E., and others (Eds.). *The Social Production of Scientific Knowledge*. Hingham, Mass.: Reidel, 1977.

Merton, R. K. "The Sociology of Science: An Episodic Memoir." In R. K. Merton and J. Gaston (Eds.), *The Sociology of Science in Europe*. Carbondale: Southern Illinois University Press, 1977.

Merton, R. K., and Gaston, J. (Eds.). *The Sociology of Science in Europe*. Carbondale: Southern Illinois University Press, 1977.

Stehr, N. and König, R. (Eds.). *Wissenschaftssoziologie*. [*The Sociology of Science*]. Opladen, West Germany: Westdeutscher Verlag, 1975.

Whitley, R. D. (Ed.). *Social Processes of Scientific Development*. London: Routledge & Kegan Paul, 1974.

1

Barbara F. Reskin

Sex Differentiation and the Social Organization of Science

In the last half of the 1960s, in response to interest in the position of women in American society, reports of the experience of female scientists began appearing in scientific journals (Rossi, 1965a, 1965b; Roe, 1966; Parish and Block, 1968; Graham, 1970; Maccoby, 1970; White, 1970). Drawing heavily on the authors' personal experience and observations, these papers introduced the status of women scientists as a subject for scientific investigation and proposed several hypotheses to account for their status. Empirical studies followed, and by 1975 researchers had accumulated considerable evidence for discrepancies in the positions of women and men in science (Simon, Clark, and Galway, 1967; Bayer and Astin, 1968; Fidell, 1970; Perrucci, 1970; Cole, 1971; Lewin and Duchan, 1971; Astin and Bayer, 1972; Reskin, 1973; Bayer and Astin, 1975) as well as in several individual disciplines (Morlock, 1973; Green,

Note: I am particularly grateful to Lowell Hargens, Joan Huber, and Paula Hudis for their extensive comments on an earlier draft of this paper and, more generally, for their valuable colleagueship. I also wish to thank Allen D. Grimshaw, Margaret W. Rossiter, and Judy Franz for their helpful comments and Carolyn J. Mullins for editorial assistance.

1976). The current state of the knowledge is summarized in two review essays (Zuckerman and Cole, 1975; Vetter, 1976) and in a review of the literature on sex differences among graduate students (Solmon, 1976).

During this period, several researchers considered whether the observed discrepancies resulted from sex discrimination, but the term *discrimination* rarely appeared in titles, and the debate was carried out circumspectly in the articles' concluding pages (see, for example, Cole and Cole, 1973; Zuckerman and Cole, 1975; Reskin, 1976). Recently some researchers have tried either to identify data sets (Cartter and Ruhter, 1975) or to develop statistical techniques (Reskin, 1977a) better suited to assessing sex discrimination, but much remains to be done on both tasks.

Thus, the study of sex differences in science has become a quasi-specialty, but work in this area has rarely been specific to science. Instead, it has depended primarily on more general theories developed to explain sex differences among academics or other professionals (Epstein, 1970a). A few studies, framed in terms of the normative structure of science, have explored the extent to which status attainment within science is universalistic (Cole and Cole, 1973; Reskin, 1976). In general, however, insights from the sociology of science have contributed little to this research area, and results have offered little in return to help explain the organization of the scientific community. This is not surprising. Political rather than theoretical considerations originally prompted these studies, and a certain amount of research was necessary just to determine the actual extent of sex differences.

Now we know more about the relative positions and performance of female and male scientists (Astin, 1977; Reskin, 1978), and it is time to consider how the organization of scientific work affects the presence, form, and extent of sex differentiation in science. For example, what aspects of the social organization of science foster sex differentiation? What aspects impede it? In what way are social patterns of sex-role differentiation reproduced in the scientific community? How does sex differentiation affect role performance?

In this chapter I consider some ways in which specific features of the scientific enterprise intereact with social patterns of sex differentiation and stratification to introduce sex differentiation into the scientific community. My discussion is necessarily speculative. We know too little about actual patterns of sex differentiation in collegial networks, patterns of collaboration, or the functional division of labor to make firm claims. At the very least, this chapter should illustrate the value of grounding future research on sex differences in our knowledge of the social organization of the scientific community.

Scientific Colleagueship and Sex Differentiation in Science

Scientists advance knowledge, train future scientists, and administer organizations that fund or pursue research (Zuckerman and Merton, 1972, p. 314). Of these activities, teaching and administration are necessary for the practice of science, but without research science could not exist. To make research possible, the research role prescribes a set of technical behaviors and an orientation with which scientists approach their own work and that of others (Feigl, 1953, pp. 8–18; Merton, 1957; Storer, 1966, p. 81). Because research occurs in a social context, the role of researcher specifies certain reciprocal relations with other researchers. These may occur as public, formal exchanges, such as presentations at meetings or published reports and the responses they elicit (Hagstrom, 1965; Storer, 1966), or as the informal interaction that comprises most exchanges between scientific workers. These reciprocal relations include the collegial behaviors that are implicit in the research role, and they, too, are necessary for carrying out scientific research.[1]

Some scientific roles—for instance, teacher and apprentice, project director and research associate—involve partners of unequal status. In contrast, the collegial role implies an egalitarian, symmetrical relationship between scientific researchers. The norms of science dictate some aspects of the collegial role. For example, communalism requires scientists to share their findings with coworkers, who in turn are obliged to evaluate them critically (Storer, 1966, p. 90). But scientists cannot follow even a fraction of the papers in their field (Price, 1963), much less respond competently to them all; so, realistically, collegial obligations are limited primarily to institutional and specialty colleagues, among whom personal contact is possible.

Other aspects of the collegial role have developed in response to the structure and needs of the scientific enterprise. The size and the complexity of modern science make it impossible for scientists to monitor all developments in their field, so they rely on colleagues to call to their attention references or results that they may have missed. Similarly, the high degree of specialization encourages scientists to draw upon colleagues for technical assistance, occasional instruction, and collaboration. The difficulty and uncertainty of research make scientists dependent on colleagues for advice and encouragement when they get hung up in their work; as potential competitors, colleagues provide the impetus for completing and publishing work. The reward system directs scientists to colleagues for appraisal, recognition, and criticism of their work. Because clearly defined standards of performance seldom exist, colleagues function as a reference group against whom scientists can evaluate their own performance. And, finally, the long hours associated with scientific work often limit scientists' opportunities to develop primary relationships away from their work, so they turn to colleagues for social companionship. In sum, the collegial role calls for scientists to teach, collaborate with, encourage, inform, evaluate and reward, compete with and befriend co-workers. Such reciprocal relations are essential for the pursuit of scientific inquiry.

Nothing in the collegial role inherently precludes collegiality between female and male co-workers. However, since the role applies to relationships between scientific researchers of approximately equal status, collegial responses usually do not evolve naturally between researchers of unequal status on either scientific or nonscientific dimensions. Instead, such co-workers usually relate to each other in terms of roles that do not assume equality. For example, few scientists treat their students and technicians as colleagues; their statuses do not elicit collegial responses, and the scientific roles of teacher-student (or administrator-employee) provide appropriate alternative models for interaction. Similarly, because women's lower gender status is inconsistent with the implicit status equivalence of colleagues, sex stratification itself can block normal collegiality between male and female researchers.

Extrascientific status distinctions also may hinder normal collegial relations in other ways. Certain "master status-determining traits" (Hughes, 1945, p. 357; or simply "master statuses," in Kanter's, 1977, terms)—such as sex, race, and age—typically take precedence over other categories in dictating social responses. These statuses are usually elaborated in and supported by a variety of role relationships that specify modes for interaction between role partners who occupy different positions on the master status dimension. In the case of gender, such role relationships include wife-husband, mother-son, father-daughter, boyfriend-girlfriend, sister-brother, lovers, and so forth. Specific role requirements for each vary across social groups, and role partners have considerable latitude. But most Americans, including scientists (Mitroff, Jacob, and Moore, 1977, p. 323), endorse and conform to the sex roles appropriate to their social positions. Whereas collegial roles are symmetrical and egalitarian, sex roles are differentiated by gender and in most cases hierarchical, granting males greater power and privilege.

Thus, female and male co-workers are subject to conflicting, mutually exclusive role models for their relationships. At one extreme, they may ignore the roles dictated by their gender status and follow the traditional collegial model. At the other, they may relate to each other exclusively in terms of the nonscientific sex roles that most closely fit their scientific situation (for example, roles primarily dictated by gender etiquette that govern the relations between female and male *non*scientific co-workers). A third alternative is minimizing their interaction. Several observers contend that the third alternative occurs frequently (Epstein, 1970a; White, 1970; Patterson, 1971, p. 233), and evidence that women are excluded from collegial networks (Bayer, 1970, p. 28; Kaufman, 1978) supports their claim. Perrucci (quoted in White, 1970, p. 414) notes that female scientists are more likely than males to stress the importance of working with colleagues and associating with scientists of recognized ability, probably because such opportunities are scarcer for them than for men. I argue, however, that male and female scientists often choose an intermediate option: they adapt conventional sex roles to the scientific setting to create a hybrid of gender and collegial roles that systematically introduces sex-role differentiation into the scientific community. Before elaborating on this argument, I should note the following qualifications.

First, the role of scientific colleague evolved when all scientists were men. The traditional masculinity of the collegial relationships probably is not sufficient to preclude cross-gender colleagueship, but it does reinforce the other barriers described above.[2]

Second, status inequality that originates outside the scientific community also can hinder the development of collegial relationships between other subgroup members. Any pronounced discrepancy in the status of co-workers could interfere with the development of a collegial relationship. If differentiated roles are associated with the status on which persons differ (as with gender status and sex roles), these may be substituted for collegial roles. For example, age, social class, and race are hierarchically differentiated, master status-determining characteristics, just as gender is; but all are not supported to the same degree by a system of status-specific roles. Age is most like gender in the existence of generally recognized roles that govern interaction between persons of different ages. Though few scientific researchers are extremely young, the youth of the rare prodigy, like the gender of a female scientist, is inconsistent with the collegial mode of interaction. Despite even prodigious accomplishments, the role responses elicited by youth are at odds with the collegial role: older scientists may behave paternalistically, while the young may respond with the deference traditionally due to older people. (See Zuckerman and Merton, 1972, for a discussion of age stratification in science.) Class and racial roles exist, but they are not universal in our society. It should not be surprising, then, that—at least with respect to class—the effects on scientific interaction are unclear (see Crane, 1969, 1976; Gaston, Wolinsky, and Bohleber, 1976).

Third, I do *not* claim that male scientists deliberately respond to female co-workers in terms of sex roles rather than collegial ones in order to hinder women's careers (although some may do so) or even that they unconsciously initiate such substitutions. Rather, the possibility of substitution inheres in the incompatibility of egalitarian collegial roles and stratified gender roles, and strict reliance on collegial roles can occur only when *both parties* treat gender as irrelevant. If scientists of either sex find a collegial relationship uncomfortable or inappropriate, they can abstain; without two willing role partners, colleagueship cannot occur. Status discrepancies potentially complicate what otherwise would be routine interaction, and for this reason alone scientists of either sex might prefer sex roles to renegotiating the nature of their relationship for each encounter (Epstein, 1970b, p. 978). Moreover, scientists learn sex roles long before they learn how colleagues interact, and some scientists of both sexes might find sex roles easier to perform. Yet women stand to lose considerably more than men do from the incorporation of sex roles into scientific interaction. Because men constitute a substantial majority in most fields, male scientists have many potential colleagues. But women who choose to relate to male co-workers in terms of modified sex roles may forgo the possibility of true colleagueship. In addition, because the content of sex roles reflects gender stratification, their use introduces both gender-based task differentiation and sex stratification into the scientific community.

Finally, some female scientists do become more integrated over time, as co-workers get used to their presence and recognize the value of their work (Pour-El, 1974, p. 37). Unfortunately, many women do not overcome their early collegial isolation and remain sufficiently productive to enjoy the ultimate rewards of collegial acceptance.

Models for Cross-Gender Relationships

If traditional sex roles provide models for relationships between female and male scientists, it follows that scientists must draw on a limited repertoire of models. Those models are based primarily on kinship or romantic relationships, although quasi-scientific superior-subordinate models (such as scientist and secretary or assistant) are also possible.

Models Based on Kinship Roles. Kinship roles provide scientists of both sexes with models for relating to co-workers. The sponsor-student relationship is the most obvious example of a model based on kinship roles; hence, we refer to scientists' "academic paternity" (in German doctoral sponsor is *Doktorvater,* or doctor-father). Some scientists construct elaborate kinship systems, tracing their academic lineage to famous scientific progenitors. Students from different cohorts who worked with the same sponsor may play mutually supportive roles, such as those of older and younger siblings or cousins (while students in the same cohort may exhibit "sibling rivalry"). These links provide neophyte scientists with a ready-made set of extrainstitutional colleagues in their specialty and can be beneficial—as long as they do not prevent the development of normal collegial relationships. And that, of course, is the crux of the issue: Do sex roles borrowed from nonscientific arenas replace or distort the normal colleagueship that is necessary for conducting scientific work? And, if they do, to what effect?

Consider the father-daughter model, in which a senior male scientist relates to a junior female paternalistically, and she responds with a combination of deference, docility, and affection. Scientists may use this model whenever an age or status difference exists, but it is most apparent in the relationships between women and their sponsors. In their broad outline, the relationships between sponsors and their protégés probably do not differ substantially by sex; both women and men profit from sponsorship in several ways. Sponsors both literally and figuratively place their students in the scientific stratification system and can intervene to provide a variety of professional opportunities. Active sponsorship also may provide young scientists with a ready-made academic extended family. These secondary connections with their sponsors' other students may be especially valuable to female scientists, who might otherwise encounter obstacles to forming collegial ties, and probably represent a large proportion of women's extrainstitutional linkages. Of course, some scientists of both sexes may dislike the paternalistic aspects of sponsorship (and women probably get larger doses of paternalism), but its advantages (Reskin, 1977b) usually outweigh any psychological costs.

The similarity in broad outline does not suggest an absence of sex differences in sponsorship. Just as father-son and father-daughter relationships differ in character, so too will relationships modeled after them. A male scientist and his sponsor both recognize that the former usually can expect eventually to be the latter's equal—even if the student must rebel to achieve that status. This usually is not true of the relationships between most female students and their male sponsors. A long-term father-daughter association is perfectly consistent with the lower status that comes with being female, and women can get trapped in long-term paternalistic relationships that discourage their scientific growth and independence. Despite the disadvantages that paternalistic sponsorship might hold, though, female scientists probably suffer more from the absence of sponsorship (Bayer, 1970, p. 13; Epstein, 1970b; American Association for the Advancement of Science, 1977) than from sex differentiation in its form.

Relationships based on the mother-son model are rare because few women scientists have access to graduate students (Bayer, 1973, p. 24), and fewer still are high enough in status to attract young male scientists. Nobel Laureate Rosalyn Yalow is among the few women scientists who successfully attracted younger male colleagues (whom she calls her "professional children"; Stone, 1978). Superficially, this model is benign (particularly for Nobel-caliber scientists), but Kanter (1977, p. 982) has noted some adverse effects on women's task performance. Female sales trainees in a predominantly male organization whose relations with male trainees were of this type were rewarded for maternal solicitude rather than independent action; as "accepting, good mothers" they could not criticize others; and they were defined as emotional rather than task specialists. If female scientists who relate to men via this role have higher status than their male role partners and are of demonstrated professional competence, these difficulties should be negligible. But if they are junior in status and are assigned to or assume the role on the basis of their age or personality, the problems that Kanter enumerates could occur.

Models Based on Marital Roles. As the most common adult sex roles, marital roles provide an obvious model for female-male relationships in other settings (Strodtbeck and Mann, 1956). Yalow's long-standing collaboration with the late Solomon Berson exemplified this pattern; although they were not married to each other, she and Berson have been likened to "an old married couple" (Stone, 1978). This model is most apt for researchers of approximately equal scientific status who are friends as well as co-workers. Thus far, because of the differences in the scientific statuses of female and male scientists, relatively few women have been professionally eligible for such a relationship. As the larger recent cohorts of female science students grow older, however, the number of such relationships may increase. Although some scientists may shun a "professional marriage" in the belief that such an intimate working relationship may compromise them, its occurrence is favored by the presence of actual, and occasionally highly visible, husband-wife teams in the scientific community (three of the five female Nobel Laureates were joint recipients with their husbands). Most of the following discussion of the "professional marriage" model applies equally to scientists who are actually married to each other; I refer to the latter as "legal" spouses to distinguish them from "professional spouses," whose relationship is only modeled after that of spouses.

Unlike parent-progeny and sibling roles, which are not gender-specific, marital roles prescribe a sexual division of labor and rewards. Hence, as models for relationships between female and male scientists, they may lead to task differentiation in which authority and creativity are reserved for the professional husband. No systematic evidence is available on sexual division of labor in relationships between professional spouses, but several of the female doctoral chemists whom I studied (Reskin, 1973), who were actually married to other chemists, worked as research associates in their husbands' laboratories. Furthermore, for professional couples who also are legal spouses, any division of labor may well be compounded by differences in rank resulting from institutional antinepotism regulations.

Role differentiation may extend to areas other than scientific work. Yalow, for example, commented that Berson would have been very upset if she did not defer to him and recounted an occasion when they visited a scientist at his home and Berson advised her to "stay with the wives" (Stone, 1978, p. 98). What is important here is that he would ask and that, even today, the prevailing wisdom among women in male-dominated fields is to talk with co-workers' wives at social gatherings, even though they may miss some professionally valuable conversations with fellow researchers.

Task differentiation on the basis of sex and marital roles will lead directly to differentiation in the allocation of credit for joint work.[3] Because the professional husband is assumed to play the dominant role, most of the credit accrues to him. For example, although Nobel Laureate Maria Goeppert Mayer and her husband,

Joseph Mayer, jointly authored the influential *Statistical Mechanics,* many people assumed that it was largely his work (Dash, 1973, pp. 33–34). Among sociologists, men were disproportionately senior authors on papers authored with women, and the sex difference was more pronounced when the coauthor was their legal wife (Wilkie and Allen, 1975, p. 22). Of course, these differences may result from the men's (husbands') higher rank (itself a reflection of sex stratification[4]), but they also are consistent with sex differentiation in the distribution of recognition.

Introducing the roles of spouses into the research setting gives rise to yet another nonscientific model for relationships between male and female co-workers. The interaction between one member of a professional couple and her or his co-workers may center around the scientific interests of the former's spouse rather than her or his own work. For professional wives, who are more likely than professional husbands to experience this, the focus of the interaction with co-workers would be their husbands' work. Even if professional husbands occasionally have the same experience (usually in interactions with their professional wives' female colleagues), it will not hamper their collegial integration, because these interactions would constitute a small portion of their total professional contacts.[5] Also, because the role of husband does not define men in the way that women are defined by being wives, occasional responses to husbands directed at their status as husbands do not preclude their simultaneously being regarded as colleagues.

The tendency for co-workers to interact with women who are either professionally or legally married to other scientists on the basis of their status as wives may lead such women to be superficially better integrated than unmarried women, but probably will not improve their chances of developing collegial relations based on their own professional interests or their status as scientists in their own right.

Regardless of their husbands' occupations, married female scientists apparently interact more frequently with male co-workers than do single women (Kaufman, 1978, pp. 16–17), perhaps because some men are more comfortable relating to married women (Bernard, 1964, pp. 198–199). For example, a senior male academic commented on his close professional relationship with a younger woman in his department (Kaufman, 1978, p. 16): "We have worked on several projects together. I find her thorough, hardworking, and very insightful . . . . I get along with her husband well too, and that makes for a good relationship as couples. That's helpful, too."

Possibly the best situation for a female scientist is marriage to a professional in another discipline. Her marital status would facilitate her social and professional integration, and the disciplinary difference would reduce the chance of her husband's receiving credit for her research contributions.

Models Based on Traditional Romantic Relationships. Social relationships characterized by some degree of sexual attraction provide additional models for cross-gender relationships between coworkers. These models, ranging from mild flirtation to love affairs, are derived from traditional patterns of romantic relationships; therefore, the behaviors involved are familiar to most adult women and men.

Flirtation is the most common of these alternatives, and for those who have trouble taking a female scientist seriously, it may make initial interaction easier. In keeping with general social patterns, women may flirt in exchange for the collegial attention they do not otherwise receive, or men may be more attentive to female co-workers who engage in some mild flirtation. At the other extreme are actual sexual relationships. Although gossip about real or suspected sexual liaisons is seldom absent when men and women work together, mentioning such relationships publicly, much less studying them systematically, has been taboo until so recently that evidence of their frequency does not yet exist. (For a theoretical discussion of this topic, see Bradford, Sargent, and Sprague, 1975.) Women scientists have only just begun to acknowledge these relationships and their structural implications—probably a first step to systematic study.

The use of traditional romantic relationships as models for cross-gender relations between co-workers adversely affects women scientists without providing any compensating advantages. For example, playing the role of girlfriend or lover usually does not improve social integration even superficially. In addition, some scientists may find the roles of rational scientist and sexually attractive woman incompatible (Bernard, 1964, p. 198). Thus, women who are sexually attractive may have trouble shifting into the role of scientist, and some scientists may be unwilling to grant it to women who relate to male co-workers in terms of a traditional romantic model, even if the men themselves initiated that pattern of interaction.

Apart from the other disadvantages of these models, the sexual double standard holds the female partner responsible for romantic or sexual involvements with male co-workers. Although women customarily retain the right to accept or decline romantic overtures, male scientists probably have considerably more leeway in choosing which models to use and can avoid traditional romantic models altogether. Women, especially younger ones, may have more difficulty avoiding these models, and those who successfully eliminate all sexual or romantic aspects from their interaction do so at a price. They may try to play the mother or daughter roles discussed above (and accept the associated costs), or they may try to interact in a wholly professional way, but many of their associates will interpret this behavior as cold and unfeminine. Rosalind Franklin's appearance apparently evokes such a reaction from James Watson (Sayre, 1975, p. 21). Kanter presents a general discussion of the fate of the "iron maiden" in male-dominated organizations.

Because romantic roles depart from strict professional roles, even the suspicion of a sexual alliance casts doubt on whether women's professional achievements result from their own merit rather than from sexual favoritism (Rossi, 1973, p. 513). Consequently, even women who would never dream of sexual involvement with a co-worker monitor their behavior to avoid the appearance of trading on their sexual attractiveness. For example, to stop gossip by co-workers, a female informant had to quit traveling with her male assistants on research trips.

Some scientists believe that sexual attractiveness gives women a professional advantage and understandably resent the possibility that a co-worker will be rewarded on grounds other than performance, especially when the channels for those rewards are not available to them. However, the advantages that a few women might obtain are small and fleeting; and, in the long run those advantages will prevent their being treated as serious colleagues or given full credit for their achievements. True collegial relations between persons who are unequal on some nonscientific status dimension can occur only if the discrepant status dimension is ignored altogether rather than emphasized through status-differentiated behavior.

Models Based on Quasi-Scientific Roles. Two additional models for relationships between female and male scientists—the scientist-technician and the scientist-apprentice models—can be considered together. In neither case are the roles gender specific, but women in research settings disproportionately hold the lower-status positions of technician or student. Because these roles are well known to all scientists and readily available as models for relationships between female and male researchers, they may be more acceptable than models based on sex roles. However, both models put women into subordinate roles incommensurate with their professional qualifications. (Professional women in other occupations also are vulnerable to role demotion, and some respond by remaining ignorant of the skills associated with the traditional female roles to which they might be demoted by co-workers; one example is the female manager who assiduously avoids learning to type.) No systematic evidence exists for the frequency of such models among female and male researchers with equivalent credentials, but anecdotal evidence documents their occurrence. For example, Nobel Prize winner James Watson (1968, pp. 20–21) apparently believed that Rosalind Franklin was co-Laureate Maurice Wilkins' assistant, although this was not the case and neither Wilkins nor the director of the Kings College Laboratory ever represented her as such (Sayre, 1975, p. 20). At least Watson was in good company, for Linus Pauling made the same mistake (Hubbard, 1976, p. 234). Sayre's (1975) account of Franklin's role in the discovery of the structure of DNA suggests that this misidentification helped Watson to justify using Franklin's crystallographic data without her knowledge or permission.

The features of female scientists' careers encourage the substitution of technician or student roles for that of colleague. First, women are disproportionately inbred at their Ph.D. departments

(Reskin, 1973), where they previously related to faculty in the professor-student model. Second, women doctorates in science who work at universities are disproportionately employed as research associates. Given the ambiguity of the research associate role, an incumbent—even one who holds a Ph. D. degree—is particularly likely to be defined as a technician rather than as a full professional. The roles of both student and technician are characterized by lower status and by a technical division of labor that allocates scientific creativity and decision making to scientists and laboratory work to those assigned the role of technician or student. Credit, presumably, is distributed accordingly. Both roles also foster female incumbents' structural dependence on higher-status males for direction and approval, thereby discouraging in the women the independence and creativity that are necessary for professional advancement.

Sex Differentiation and Scientific Processes

Space does not permit a full analysis of the implications of sex differentiation for women's participation in the social organization of science. Therefore, I have selected as illustrations two processes that are particularly crucial to the research role: informal communication and patterns of collaboration.

Informal Communication. The communication system in science includes all institutional arrangements and customs that affect the transmission of scientific messages among scientists (Menzel, 1962, p. 419). Upon discovering the importance to researchers of apparently accidental information, Menzel concluded that informal information networks actually constitute a highly organized communication system. Similarly, in a 1959 study Glass and Norwood (cited in Kjerulff and Blood, 1973, p. 625) found that casual conversation is the most important source of research information. Scientists use informal communication to supplement their knowledge of findings and to obtain technical information and procedural details unavailable in the literature. Colleagues provide information about unpublished work or about published work that a co-worker may not have read. Informal discussions also can reveal common interests, provide scientists with collegial reactions, and may even lead to collaboration. They may also alert scientists to the risk of being anticipated and can help to ensure the priority of unpublished work.

A variety of professional (as distinguished from scientific) information also is transmitted informally: sources of materials, data, or services; tips about funding opportunities, potential employees, or job prospects; and news of the current research of other scientists. Although this information is sometimes available through formal channels, acquiring it can occupy most of a scientist's time. Moreover, some professional gossip may never become available outside informal networks. Missing out on stories about a colleague's squash game or prodigious drinking capacity will not hamper a scientist's performance, but careers can suffer if a scientist does not hear gossip about who is helpful and who is not to be trusted, or who is an operator and who is a "son of a bitch." Hagstrom (1965, p. 115) quotes an informant who was warned, on his first job, against working with an older man who exploited collaborators. Information about other scientists' moves, collaborations, or memberships on review panels also is useful.

Information flows into a local system through its members' contacts with outsiders and is transmitted within it during daily shoptalk in the labs and corridors and at lunch, in the locker room, over drinks, or during the evening while their families socialize. Scientists not present for such conversations miss out on their content. Conversations held away from the work setting obviously exclude persons not involved in such activities.

The exclusion of women from conventional collegial relationships, the substitution of gender-based role relations for collegial roles, and other forms of sex-role differentiation all reduce women's access to informal communication. Sex-role differentiation will reduce the probability that male scientists will define a woman as an interested and trustworthy recipient for various types of information (Hughes, 1945, p. 356) and even the likelihood that she will be present when information is being exchanged. Scientists may not remember to tell women about funding opportunities or special conferences if the women's roles with male co-workers are modeled after that of the student or technician. If it occurs to them at all, they probably assume that women with close ties to a particular man will receive the information from him. Even if that assumption is sound, forgetfulness and delays in the transfer of information may put the woman at a disadvantage. The ultimate result is that any marginality in women's positions in informal communication systems is self-perpetuating: women receive less information and hence lack information to exchange, and so on. Also, women's assignment to or acceptance of social-emotional roles such as the mother role could lead them to specialize in personal rather than professional information, thereby reinforcing their image as oriented toward people rather than the profession.

Extrainstitutional communication flows between former students and teachers, former co-workers, and scientists who became acquainted at professional meetings, during laboratory visits, or through correspondence. However, because female faculty teach fewer graduate students (Patterson, 1971; Bayer, 1973, p. 24), they have fewer former students to transmit information. Because their ties with their academic sponsors are weaker (Roe, 1966; Bayer, 1970, p. 13), they are less likely to be in touch with them. If their relationships with former co-workers were not of the traditional collegial variety, they probably did not persist when one party moved. When it comes to contacts outside their institution (Bayer, 1970, p. 18), female scientists are especially disadvantaged. Female bioscientists received fewer invitations to participate in events away from their labs (Bernard, 1964; Bayer, 1973, p. 28) and received fewer preprints from scientists at other institutions (Bernard, 1964).

When women do interact with noninstitutional colleagues at meetings, their interactions may follow traditional sex roles (Mintz, 1967) and thus have little professional value. Female bioscientists found discussions at meetings a less important source of scientific information than did males, and women zoologists were substantially less likely to report that their most productive conversations were with persons whose work was just becoming known (Bernard, 1964). Because scientists find informal conversations more useful than the formal presentations (Kaplan and Storer, 1968, p. 115), women's restricted access to such conversations could explain these differences. Male scientists who are unwilling to extend colleagueship to a woman also will be constrained in dealing with female scientists whom they do not know and may refrain from approaching a woman after she has given a formal presentation. If male scientists typically hesitate about such contacts, then women will lack feedback on presentations that other scientists find extremely valuable (Lin, Garvey, and Nelson, 1970, pp. 30–31).

Scientists who are excluded from informal communication networks depend on less efficient and more time-consuming formal channels of information. For example, young people—presumably not yet well integrated in informal communication networks—were overrepresented among users of an innovative system that distributed unpublished material (Griffith and Miller, 1970, p. 135). Women bioscientists, predictably, depended more than men on published literature (Kaplan and Storer, 1968, p. 115).

Unlike formal communication, informal communication is a two-way process that helps researchers establish the relevance of others' results for their own work, direct the communication process to their particular needs, speculate about findings, and obtain immediate feedback (Garvey and Griffith, 1967, p. 1013). But two-way communication is less likely to occur between female and male co-workers whose relationship is modeled after roles that assign initiative to the male and define the female as the receiver of messages (information, directions, and so forth; see Mintz's 1967 account of the typical communication between female and male sociologists at their annual professional meetings). For women, then, sex differentiation reduces the value of informal communication by reducing the likelihood of two-way communication.

Collaboration. Patterns of collaboration among scientists reflect work organization, scientific stratification, specialization, the cognitive structure of specialties, and informal collegial relations as well as the preference of individual scientists for joint or solo work. Because collaboration patterns are influenced by both formal and informal aspects of the social organization of science, they too will reflect sex differentiation in the scientific community. Hagstrom (1965) provides several insights that suggest hypotheses about sex differentiation in collaboration. Because collaboration usually begins informally, anything that affects the frequency of informal contacts between colleagues will affect the likelihood of collaboration (Hagstrom, 1965, p. 114 and p. 122, table 12). Scientists broach the idea of collaboration cautiously, particularly if the suggestion might imply their inability to solve a problem independently. They can reduce the risk of being turned down by tentatively sounding out a potential collaborator in advance, but the possibility of rejection probably deters individuals who are particularly vulnerable. For this reason and others, collaboration tends to occur among equals (Hagstrom, 1965).

Sex-role differentiation in science may affect female scientists' opportunity to collaborate. If it restricts their participation in informal communications systems, it will reduce their access to potential collaborators. Certain borrowed models of relationships will hinder collaboration. Women in explicitly lower-status positions than those of their male role partners (the apprentice, technician, or daughter roles) are more vulnerable to rejection and should be more reluctant to propose collaboration. Although male scientists are not constrained from initiating collaboration, the tendency for collaborators to be peers should lead joint efforts among unequals to exhibit a substantial division of labor by gender, assigning decision making and creative tasks to the male. Moreover, if the women's lower status in such relationships is generally known, other co-workers might refrain from proposing collaborative work. To propose collaboration might imply their own professional inferiority to the women (and, conceivably, by transitivity, to her male role partners as well).

Roles that link female scientists with particular male colleagues also might discourage other scientists from viewing them as potential collaborators, especially if exclusive access is implied in the borrowed roles as defined in the larger society.[6] If this reasoning is correct, proportionately more women than men should collaborate primarily with a single colleague rather than with several different collaborators. Female researchers who are legally or professionally married to other scientists should be freer than other women to initiate collaboration, both with their professional spouses and with other male scientists.

Hagstrom (1965, p. 118) suggests that scientists are less reticent about asking someone with obvious expertise to collaborate, because the request does not imply their own deficiency. So women who are clearly experts in their specialties may have more opportunities to collaborate (Hagstrom, 1965, p. 124; Epstein, 1970b, p. 979). However, since these collaborative relationships would not be between equals, they probably would be limited to the duration of the project, and the interaction they involved might be relatively formal. For these reasons they would not necessarily improve the female researcher's integration into the collegial system or even increase her future chances to collaborate. In fact, they may help to define her as a technical expert rather than a full colleague.

Other variables also may reduce collaboration between the sexes. Scientists who believe that male-female working relationships have a sexual component would avoid working with women lest others misinterpret the association. Although in jest, Yalow's dissertation director explained that he accepted her as a graduate student because he was so old that his motives would not be suspect (Stone, 1978, p. 30), thus illustrating how near the surface such concerns often are. Some scientists may simply feel uncomfortable working with members of the opposite sex. Male researchers might have such feelings for a whole host of reasons related to women's discrepant gender status. Women may not pursue or may even avoid opportunities to collaborate with men because they fear—with some reason—that they will not receive full credit for their contributions. Women who work with their legal husbands are particularly vulnerable in this regard; even first authorship for them may be discounted as an expression of conjugal affection (Dash, 1973, p. 289). The evidence on sex differences in frequency of collaboration is largely anecdotal, though recent data show that female sociologists are less likely than males to publish collaborative papers (Chubin, 1974, p. 85; Mackie, 1976, p. 285). The magnitude of the differences should vary with the proportion of workers in a field who are female, but data from other disciplines are not yet available.

Indirect evidence of sex differentiation in science should exist in patterns of name order on coauthored papers. Conventionally, first authorship denotes greater contribution to the research and customarily goes to the scientist who initiated the collaboration (Hagstrom, 1965, p. 117), although it also may mirror differences in the academic statuses of the authors. The order in which scientists' names appear on multiauthored papers is of consequence because recognition and other rewards often depend on the relative contribution that name order denotes (Zuckerman, 1968). If sex differentiation discourages women from proposing collaborations, it would indirectly reduce their likelihood of being senior author on joint papers (Wilkie and Allen, 1975, p. 22). The specific roles adapted for female-male relationships between collaborators also may affect the division of responsibility and, thus, credit for joint work. Both the scientist-technician and the professor-student relationships, which the sexes may use as models, routinely assign first authorship to the higher-status party—invariably the male. Men in these positions have the prerogative of dictating a different name order, but they also have the prerogative of omitting the names of junior collaborators. Mackie (1976, p. 289) speculates that female collaborators, like graduate students or technicians, may sometimes receive acknowledgments rather than coauthorship in return for their contributions. In addition, the division of labor congruent with borrowed roles may discourage women from contributing to the more creative aspects of the joint ventures, in which case they would not merit first authorship.

Sex Differentiation and Women's Scientific Performance

Sex-based functional differentiation in science has been well documented. Female and male scientists tend to work in different kinds of organizations and, within those organizations, typically fill different positions and perform different functions (Vetter and Babco, 1975; Gilford and Snyder, 1977). Some observers have suggested that such gender-based functional differentiation accounts in part for women's lower productivity. (Solmon, 1976, p. 24; Astin, 1977, p. 5). Sex differences are reduced though not eliminated by controlling for employment sector (Centra, 1974, pp. 68–72). More detailed examination of the actual roles female and male scientists occupy should further reduce these differentials. Consider the results of a survey of college and university faculty by the American Council on Education (Bayer, 1970, 1973). Female university faculty members spent more time than males teaching and advising students and correspondingly less time in administration and research. They taught more classes (including more introductory courses), were in class more hours per week, and not surprisingly were less satisfied with their teaching load. Although they taught more undergraduates, they were less likely to have teaching assistants. They also taught fewer graduate students and had fewer research assistants than did male faculty. That substantially more men than women held Ph.D. degrees accounts for some of these differences, and comparisons within fields should reduce the differences still further. Unfortunately, available information on the principal activities of university scientists is not presented separately for both field and sex (Vetter, 1977, pp. 24–25). But the differences reviewed here probably do not stem entirely from disciplinary differences. More likely they reflect real sex differences in role obligations within the sciences.

It is easy to see how these differences could affect performance of the research role. Undergraduate teaching takes time away from research without providing an opportunity to recruit students or get feedback on one's work. Bayer (1970) reports that university faculty women are more likely than men to agree that they hardly ever have time to give a piece of work the attention it deserves.

Teaching graduate students in one's area facilitates research. Seminars provide an opportunity for researchers to discuss their ongoing work with an intelligent captive audience and to attract students to a research project. If women teach fewer graduate students, they will find it more difficult to recruit them. In the experimental sciences, where research is a team effort, having fewer students presumably reduces women's research output. Hagstrom (1965, p. 120) found a negative correlation between the number of collaborators that scientists had and their number of graduate students (and both are correlates of productivity). Scientists with weaker collegial ties can compensate by working with a large group of students. Ironically, women apparently do not have this option; both their sex and their isolation from other members of the department probably deter the better students from working with them.

As mentioned, collegial ties and informal communication also are essential for scientific research (Crane, 1969; Mullins, 1973; Hargens, Mullins, and Hecht, 1977). Hagstrom (1965, p. 50) reports that the number of papers his informants published in the previous three years was moderately related to their communications with departmental colleagues and strongly related to communications with disciplinary colleagues at other departments. Pelz (1956) found that the most productive medical researchers working in government typically had one close colleague with similar values and had frequent contacts with several additional colleagues. If colleagueship is causally prior, then these associations point to obstacles to women's scientific performance. Lack of communication with colleagues limits access to necessary information, and women are more poorly integrated into informal communication systems. Thus, the "lucky accidents" that male biochemists experienced in their research were more likely to result from informal communication than those the women experienced. (Bernard, 1964, p. 273). Restricted communication also reduces scientists' motivation to do research (Hagstrom, 1965, p. 48–49), partly because poorly integrated scientists receive less encouragement and fewer informal rewards for their achievements.[7] The stronger impact of formal citations on the productivity of women chemists (Reskin, 1978) is consistent with their receiving less informal recognition. Colleagues' expectations should be especially important to researchers who face conflicting demands from other institutional roles and rewards for nonresearch performance, and gender-based functional differentiation in the scientific community concentrates women in just such situations. Unfortunately, co-workers who do not expect women to be productive are unlikely to provide encouragement and informal rewards, thereby contributing to the fulfillment of their expectations.

Zuckerman and Merton (1972, p. 323) suggest that retention of the research role by older scientists depends partly on their having active researchers as their reference group. This applies equally to women, whose isolation may lead them to form reference groups among people outside their discipline or among nonscientists (teachers, administrators, or other professional women). Kaufman (1978, p. 17) found that female faculty in human ecology were more likely than male co-workers to name as members of their collegial network persons whose research interests differed widely from their own. One is hardly surprised, then, to learn that these women were less likely than the men to find their colleague-friends professionally valuable.

Sex differentiation in collegial networks will increase the likelihood that women will be anticipated in publishing their research. Scientists are not anticipated by institutional co-workers, who know each other's work, but by people at different institutions. Researchers who learn through the grapevine that they are working on the same problem can forestall anticipation by agreeing to collaborate rather than compete (Hagstrom, 1965, p. 115), but collaboration is less available to female scientists. More poorly integrated into informal communication systems, they are less likely to learn of others who are working on the same problems. If they do learn of potential competitors, they are less likely to know them than male researchers would be, because they are more isolated from noninstitutional colleagues in their specialty. Finally, women probably would be more hesitant than men about proposing collaboration with scientists whom they do not know, because they are unsure of their reception (Epstein, 1970b, p. 976). As a result, women who are active researchers probably face a greater risk of being anticipated than similarly situated men do.

Summary and Conclusions

The collegial role prescribes a set of reciprocal obligations between researchers of approximately equal status, involving the exchange of information, encouragement, evaluation, and informal recognition. Without these collegial supports, performance of the scientific role is almost impossible. Because pronounced discrepancies in the scientific or social statuses of co-workers are inconsistent with colleagueship, professional and technical workers of unequal scientific status relate in terms of other institutionalized roles. If male and female scientific co-workers do not both choose to ignore the status discrepancy introduced by gender, they cannot interact as full colleagues. Alternatively, they may minimize their interaction, thereby isolating female scientists, or they may draw on nonscientific role relationships that are consistent with their gender statuses but inconsistent, to varying degrees, with the scientific role. Either result reduces the likelihood that women will be fully integrated into collegial relations. In short, obstacles to collegiality between the sexes help to introduce sex differentiation into the scientific community. Its presence, in turn, will be manifest in the positions that the sexes typically occupy and in differences in their scientific role performance.

The evidence I have offered for this thesis is primarily anecdotal. Indeed, the argument itself is largely conjectural and should profit from the insights of other observers of the scientific community. Ultimately, a variety of factors that I could not consider here—factors such as the proportion of women in a discipline, its size and age, and its level of consensus or codification (Zuckerman and Merton, 1972)—may specify the processes of sex differentiation. But, regardless of whether each claim stands or falls, the foregoing should demonstrate the value and indeed the urgent need to go beyond simple questions about the extent of sex differences in science to examine how the social organization of scientific work and the research role foster sex differentiation in science.

Notes

1. Despite the importance of collegiality, analyses of the role of scientific researcher usually have ignored collegial obligations, probably because most scientists learn the collegial role without formal instruction and perform it automatically. Individually, we probably attribute poor colleagueship to idiosyncratic factors such as "personality," especially if the research role is performed adequately in other respects.

2. Were collegial roles to evolve in some new, sexually integrated profession, we could determine the importance of tradition relative to barriers resulting from the nature of colleagueship (although in sexually integrated professions gender also should be less salient; see Kanter, 1977). But an increased proportion of women in a profession should also lead to various mechanisms that would mitigate the strain engendered by mutually exclusive gender and collegial roles. Segregation (Hughes, 1945, p. 358; Epstein, 1970a, p. 187) and functional differentiation are two possibilities. A third—substantially improved collegial integration of female scientists—seems less likely, given the apparent incompatibility between collegial equality and gender stratification.

3. This situation holds for all borrowed roles that assign women an inferior status; indeed, it is consistent with sex-role differentiation in general.

4. Wives often hold lower professional status than their husbands for two reasons. First, in keeping with social norms that husbands should hold higher status than their wives (an arrangement that reinforces men's gender-based authority and affords their spouses' apparent upward mobility), scientists may seek partners who will conform to the expectation. (Marriages between female graduate students and male faculty, which are favored by both proximity and the social norm, ensure the husbands' professional superiority.) Second, among married couples who initially held equal professional status, the husbands may advance more rapidly for a variety of reasons, some of which are discussed in this chapter.

5. An exception would be a man married to a female scientist of such high status that most of his colleagues' interaction with him would focus primarily on her work rather than his own. Husbands of very successful women in professions such as entertainment, sports, or politics also experience this and may assume some semiprofessional role directing their wives' careers. Such a role both legitimizes their interacting with others in terms of their wives' status and permits the appearance of dominance. This option is not available to men trained in research, and the relationships of scientific couples in this situation may be strained.

6. Scientists' wariness about appropriate behavior toward their colleagues' spouses when the latter are also scientists is reflected in their tendency to accompany professional contacts with jokes about jealous spouses.

7. Epstein (1973, p. 68) has suggested that women may be rewarded for merely adequate performance, because some men find *any* performance by a woman commendable, or that they may even be rewarded for irrelevant performance. Of course, rewards that are not contingent on performance do not encourage professional achievements.

References

American Association for the Advancement of Science, Office for Women in Science. *Conference on the Participation of Women in Scientific Research.* Washington, D.C.: American Association for the Advancement of Science, 1977.

Astin, H. S. "Factors Affecting Women's Scholarly Productivity." In H. S. Astin and W. Hirsch (Eds.), *Women: A Challenge to Higher Education.* New York: Praeger, 1977.

Astin, H. S., and Bayer, A. E. "Sex Discrimination in Academe." *Educational Record,* 1972, *53,* 101–118.

Bayer, A. E. *College and University Faculty: A Statistical Description.* Vol. 5. Washington, D.C: American Council on Education, June 1970.

Bayer, A. E. *Teaching Faculty in Academe: 1972–73.* Vol. 8. Washington, D. C.: American Council on Education, August 1973.

Bayer, A. E., and Astin, H. S. "Sex Differences in Academic Rank and Salary Among Science Doctorates in Teaching." *Journal of Human Resources,* 1968, *3* (2), 191–200.

Bayer, A. E., and Astin, H. S. "Sex Differentials in the Academic Reward System." *Science,* 1975, *188,* 796–802.

Bernard, J. *Academic Women.* University Park: Pennsylvania State University Press, 1964.

Bradford, D. L., Sargent, A. G., and Sprague, M. S. "Executive Man and Woman: The Issue of Sexuality." In F. E. Gordon and M. H. Strober (Eds.), *Bringing Women into Management.* New York: McGraw-Hill, 1975.

Cartter, A. M., and Ruhter, W. E. *The Disappearance of Sex Discrimination in First Job Placement of New Ph.D.s.* Los Angeles: Higher Education Research Institute, 1975.

Centra, J. A. *Women, Men, and the Doctorate.* Princeton, N.J.: Educational Testing Service, 1974.

Chubin, D. "Sociological Manpower and Womanpower: Sex Differences in Career Patterns of Two Cohorts of American Doctoral Sociologists." *American Sociologist,* 1974, *9* (2), 83–92.

Cole, J. R. "American Men and Women of Science." Paper presented at 66th annual meeting of the American Sociological Association, Denver, Colo. Aug. 31, 1971.

Cole, J. R., and Cole, S. *Social Stratification in Science.* Chicago: University of Chicago Press, 1973.

Crane, D. "Social Class Origin and Academic Success: The Influence of Two Stratification Systems on Academic Careers." *Sociology of Education,* 1969, *42* (1), 1–17.

Crane, D. "Reply to Gaston, Wolinsky, and Bohleber." *Sociology of Education,* 1976, *49,* 187–189.

Dash, J. *A Life of One's Own.* New York: Harper & Row, 1973.

Epstein, C. *Woman's Place: Options and Limits in Professional Careers.*

Berkeley: University of California Press, 1970a.

Epstein, C. "Encountering the Male Establishment: Sex-Status Limits on Women's Careers in the Professions." *American Journal of Sociology,* 1970b, *75* (2), 965–982.

Epstein, C. "Bringing Women In." *Annals of New York Academy of Sciences,* 1973, *208,* 62–70.

Feigl, H. "The Scientific Outlook." In H. Feigel and M. Brodbeck (Eds.), *Readings in the Philosophy of Science.* New York: Appleton-Century-Crofts, 1953.

Fidell, L. S. "Empirical Verification of Sex Discrimination in Hiring Practices in Psychology." *American Psychologist,* 1970, *25,* 1094–1098.

Garvey, W. D., and Griffith, B. C. "Scientific Communication as a Social System." *Science,* 1967, *157,* 1011–1016.

Gaston, J., Wolinsky, F. D., and Bohleber, L. W. "Social Class Origin and Academic Success Revisited." *Sociology of Education,* 1976, *49,* 184–187.

Gilford, D. M., and Snyder, J. *Women and Minority Ph.D.'s in the 1970s.* Washington, D.C.: National Academy of Sciences, 1977.

Graham, P. A. "Women in Academe." *Science,* 1970, *169,* 1284–1290.

Green, Sister A. A. "Women on the Chemistry Faculties of Institutions Granting the Ph.D. in Chemistry." Paper presented at meeting of American Chemical Society, Oct. 1976.

Griffith, B., and Miller, A. J. "Networks of Informal Communication among Scientifically Productive Scientists." In C. E. Nelson and D. K. Pollock (Eds.), *Communication Among Scientists and Engineers.* Lexington, Mass.: Heath, 1970.

Hagstrom, W. O. *The Scientific Community.* New York: Basic Books, 1965.

Hargens, L. L., Mullins, N. C., and Hecht, P. K. "Research Areas and Stratification Processes in Science." Paper presented at 2nd annual meeting of the Society for the Social Studies of Science, Boston, Oct. 1977.

Hubbard, R. "Rosalind Franklin and DNA." Book review in *Signs,* 1976, *2,* 229–237.

Hughes, E. C. "Dilemmas and Contradictions of Status." *American Journal of Sociology,* 1945, *50,* 353–359.

Kanter, R. M. "Some Effects of Proportions on Group Life:

Skewed Sex Ratios and Responses to Token Women." *American Journal of Sociology,* 1977, *82,* 965–990.

Kaplan, N., and Storer, N. W. "Scientific Communication." In D. L. Sills (Ed.), *International Encyclopedia of the Social Sciences.* Vol. 14. New York: Macmillan and Free Press, 1968.

Kaufman, D. R. "Associational Ties in Academe: Some Male and Female Differences." *Sex Roles,* 1978, *4* (1), 9–21.

Kjerulff, K. H., and Blood, M. R. "A Comparison of Communication Patterns in Male and Female Graduate Students." *Journal of Higher Education,* 1973, *44,* 623–632.

Lewin, A. Y., and Duchan, L. "Women in Academia." *Science,* 1971, *173* (4000), 892-895.

Lin, N., Garvey, W. D., and Nelson, C. E. "A Study of the Communication Structure of Science." In C. E. Nelson and D. K. Pollock (Eds.), *Communication Among Scientists and Engineers.* Lexington, Mass.: Heath, 1970.

Maccoby, E. E. "Feminine Intellect and the Demands of Science." *Impact of Science on Society,* 1970, *20,* 13–28.

Mackie, M. "Professional Women's Collegial Relations and Productivity: Female Sociologists' Journal Publications, 1967 and 1973." *Sociology and Social Research,* 1976, *61* (3), 277–293.

Menzel, H. "Planned and Unplanned Scientific Communication." In B. Barber and W. Hirsch (Eds.), *The Sociology of Science.* New York: Free Press, 1962.

Merton, R.K. "Science and Democratic Social Structure." In R. K. Merton (Ed.), *Social Theory and Social Structure.* New York: Free Press, 1957.

Mintz, G. R. (pseudonym). "Some Observations on the Function of Women Sociologists at Sociology Conventions." *American Sociologist,* 1967, *2* (3), 158–159.

Mitroff, I. I., Jacob, T., and Moore, E. T. "On the Shoulders of the Spouses of Scientists." *Social Studies of Science,* 1977, 7 (3), 303–327.

Morlock, L. "Discipline Variation in the Status of Academic Women." In A. S. Rossi and A. Calderwood (Eds.), *Academic Women on the Move.* New York: Russell Sage Foundation, 1973.

Mullins, N. C. *Theory and Theory Groups in Contemporary American Sociology.* New York: Harper & Row, 1973.

Parish, J. B., and Block, J. S. "The Future of Women in Science and Engineering." *Bulletin of the Atomic Scientists,* 1968, *24,* 46–49.

Patterson, M. "Alice in Wonderland: A Study of Women Faculty in Graduate Departments of Sociology." *American Sociologist,* 1971, *6,* 226–234.

Pelz, D. C. "Some Social Factors Related to a Performance in a Research Organization." *Administrative Science Quarterly,* 1956, *1,* 310–325.

Perrucci, C. C. "Minority Status and the Pursuit of Professional Careers: Women in Science and Engineering." *Social Forces,* 1970, *49* (2), 245–259.

Pour-El, M. B. "Mathematician." In R. B. Kundsin (Ed.), *Women and Success.* New York: Morrow, 1974.

Price, D. J. de S. *Little Science, Big Science.* New York: Columbia University Press, 1963.

Radcliffe Community on Graduate Education for Women. *Graduate Education for Women.* Cambridge, Mass.: Harvard University Press, 1956.

Reskin, B. F. "Sex Differences in the Professional Life Chances of Chemists." Unpublished doctoral dissertation, University of Washington, Seattle, 1973.

Reskin, B. F. "Sex Differences in Status Attainment in Science: The Case of the Postdoctoral Fellowship." *American Sociological Review,* 1976, *41* (4), 597–612.

Reskin, B. F. "Assessing Sex Discrimination in Science." Paper presented at Symposium on Indication of Institutionalized Racism-Sexism, University of California, Los Angeles, April 1977a.

Reskin, B. F. "Academic Sponsorship and Scientific Careers." Paper presented at 2nd annual meeting of the Society for the Social Studies of Science, Boston, Oct. 1977b.

Reskin, B. F. "Scientific Productivity, Sex, and Location in the Institution of Science." *American Journal of Sociology,* 1978, *83* (5), 1235–1243.

Roe, A. "Women in Science." *Personnel and Guidance Journal,* 1966, *44,* 784–787.

Rossi, A. "Barriers to the Career Choice of Engineering, Medicine, or Science Among American Women." In J. A. Mattfield and

C. G. Van Aken (Eds.), *Women and the Scientific Professions*. Cambridge: M. I. T. Press, 1965a.

Rossi, A. "Women in Science: Why So Few?" *Science,* 1965b. *148,* 1196–1202.

Rossi, A. S. "Summary and Prospects." In A S. Rossi and A. Calderwood (Eds.), *Academic Women on the Move*. New York: Russell Sage Foundation, 1973.

Sayre, A. *Rosalind Franklin and DNA*. New York: Norton, 1975.

Schilling, C. W. "Informal Communication Among Bioscientists." Pts. 1–2. Unpublished manuscript, George Washington University, 1963–64.

Simon, R. J., Clark, S. M., and Galway, K. "The Woman Ph.D.: A Recent Profile." *Social Problems,* 1967, *15* (2), 221–236.

Solmon, L. C. *Male and Female Graduate Students: The Question of Equal Opportunity*. New York: Praeger, 1976.

Stone, E. "A Mme. Curie from the Bronx." *New York Times Magazine,* April 9, 1978, pp. 29–31, 34, 36, 95–103.

Storer, N. W. *The Social System of Science*. New York: Holt, Rinehart and Winston, 1966.

Strodtbeck, F. L., and Mann, R. D. "Sex-Role Differentiation on the Jury." *Sociometry,* 1956, *19,* 3–11.

Vetter, B. M. "Women in the Natural Sciences." *Signs,* 1976, *1* (3, pt. 1), 713–720.

Vetter, B. M. "Data on Women in Scientific Research." Unpublished paper, 1977.

Vetter, B. M., and Babco, E. L. *Professional Women and Minorities: A Manpower Data Resource Service*. Washington, D.C.: Scientific Manpower Commission, 1975.

Watson, J. D. *The Double Helix*. New York: Atheneum, 1968.

White, M. S. "Psychological and Social Barriers to Women in Science." *Science,* 1970, *170,* 413–416.

Wilkie, J. R., and Allen, I. L. "Women Sociologists and Coauthorship with Men." *American Sociologist,* 1975, *10,* 19–24.

Zuckerman, H. "Patterns of Name Ordering Among Authors of Scientific Papers: A Study of Social Symbolism and Its Ambiguity." *American Journal of Sociology,* 1968, *74* (3), 276–291.

Zuckerman, H., and Cole, J. R. 'Women in American Science." *Minerva,* 1975, *13,* 82–102.

Zuckerman, H., and Merton, R. K. "Age, Aging, and Age Structure in Science."In M. W. Riley, M. Johnson, and A. Foner (Eds.), *Aging and Society.* Vol. 3: *A Sociology of Age Stratification.* New York: Russell Sage Foundation, 1972.

2

Willie Pearson, Jr.

Race and Universalism in the Scientific Community

The norm of universalism prescribes that scientific research should be evaluated solely on its extension of knowledge about nature. Ascriptive characteristics such as sex, race, age, ethnicity, or nationality are irrelevant in the evaluation of scientists' work. It is scientists' performance, and not some particularistic criterion, that determines their position in the stratification system (Merton, 1957, 1973).

Cole and Cole (1967, 1968, 1973) report that in physics, but also in several other disciplines, the scientific community operates with much greater universalism than one suspects of other social institutions and that the American scientific community operates at a high level of universalism. Similar results were found by Gaston (1978). In a study of 300 American and 300 British scientists, he concluded that the reward system in Britain and the United States operates to a large extent in a universalistic fashion.

Sociological studies dealing with universalism in science focus on the social backgrounds of scientists—for instance, their social class origins and the colleges or universities where they took

Note: I wish to thank Sharon D. Peters and Phillip B. Middleton for their helpful comments on earlier versions of this chapter.

their undergraduate and graduate degrees. Crane (1965) suggests that American scientists' careers are influenced both by their doctoral origins and by the prestige of their current university affiliation. Scientists trained at prestigious departments are more productive than scientists from departments with lower prestige. Crane believes that scientists are recognized not solely on the basis of their scholarly contribution but partially also on the basis of their place in the social stratification of universities and university departments. To the extent that these findings are pertinent for black scientists, they suggest that productive black scientists, mostly affiliated with low-prestige historically black institutions, do not gain comparable recognition for levels of productivity similar to those of their white colleagues.

Hargens and Hagstrom (1967) indicate that ideally the reward system works like a contest. (Recognition in their paper was indexed not by honorific awards but by appointment to a prestigious department.) In reality, however, the system deviates somewhat from the contest mode and resembles more a sponsored mode—with sponsorship based on the reputation of the department where scientists received training. However, although many black scientists are trained in prestigious departments, their placement still may be affected by discrimination. That black Americans have suffered racial discrimination over a long period of American history, and that many social institutions practice institutional racism, is beyond doubt. In the educational sector, white men historically have dominated American academic employment. Some of America's most talented black scholars—such as W. E. B. Du Bois, Charles S. Johnson, and Horace Mann Bond—never held a full-time position at a major research university. In 1945 fewer than 2 percent of all blacks with a Ph.D. degree held academic appointments outside historically black colleges and universities (Freeman, 1976a, 1976b).

In contrast, recent studies show that highly educated blacks are beginning to attain some parity with their white peers, although the economic conditions of blacks in general continue to be low. Between 1969 and 1973, the proportion of black college teachers employed in predominantly white schools nearly doubled (from 29 to 49 percent), and the average academic salaries of blacks came to equal or in some instances exceed those of whites. Specifically, the salaries of black academics with five or more scholarly articles tend to be higher than those of comparable whites (Mommsen, 1974; Freeman, 1976a, 1976b).

Role Performance of Black Scientists

A thorough search of the relevant sociology of science literature reveals few studies on role performance of blacks in science. Herman R. Branson (1952, 1955), a highly productive black physicist, was one of the earliest to assess the role performance of black American scientists. Although his 1952 study was not systematic, he succeeded in identifying several productive black scientists who were publishing or who had published research papers in recognized scientific journals of international circulation, or who were working in the basic or developmental research laboratory of a major industry. Branson's study included primarily case histories on living and dead black science doctorates who had made significant contributions to their fields. Underlying Branson's paper is the implicit assumption that black scientists are not given their just reward (recognition).

Taylor, Dillard, and Proctor (1955) report an unsystematic survey of black scientists in chemistry, biology, mathematics, and physics. Of the thirty-five biologists (91.4 percent had the Ph.D. degree) responding, the average age at first major publication (defined as an article in a leading journal) was 32.1 years. The average number of major publications was 6.9. During an average of 10.8 years after receipt of the degree, the mean rate per year was 0.49 major articles and 0.46 minor articles. Their sample of fifty black chemists (68 percent had the Ph.D.) produced about 9 major career publications per individual. In their sample of twelve mathematicians (five with the Ph.D.), 48 major publications were produced by six individuals, but two persons were responsible for 41 of the 48 papers. One of those two had been affiliated during the previous twelve years with an established graduate school. Finally, their sample of seventeen black physicists (53 percent had the Ph.D.) showed that the highest rate of productivity was achieved by physicists who earned the Ph.D. before they were 30 years of age. The average number of articles per year since the Ph.D. was about 1.40.

These data are sketchy, and the authors did not attempt to examine (or at least they did not report) the same variables for each of the four disciplines. These unsystematic data cannot lead to any conclusions or reliable predictions about the level of research productivity among black scientists, or tell us whether white scientists perform differently, on the average, from black scientists.

In a study of the productivity of 485 sociology doctorates (55 blacks and 430 nonblacks), Clemente (1974) found that nonblacks published more than blacks. After controlling for other relevant variables, however, he concluded that race appears to make little or no difference in total career publications. Because of the serious shortcomings of Clemente's paper, which he describes, his study does little to extend our knowledge about the role performance of blacks in sociology.

In a comparative study of black and white political scientists, Puryear, Woodard, and Gray (1977) report that the leading study of political scientists (by Somit and Tenenhaus) completely ignored the role and status of black academics and that the American Political Science Association (APSA) in 1969 could identify only eighty black political science doctorates in America. White faculty members, compared to the blacks, tend to participate more in their professional association. Approximately 88 percent of the whites were members of the APSA, but only about two thirds of the black faculty were members. To what extent professional membership affects role performance is not shown in the study.

Some qualitative data are instructive at this point. Charles Henry Turner (1867–1923), an entomologist, who completed his Ph.D. degree at the University of Chicago in 1907, was a highly productive scientist. From 1892 to 1923, he published about forty-nine scientific papers in the major scientific journals of his day and is the discoverer of a behavior in insects (called Turner's circling). Despite the high quality of his research, he spent most of his working years at Sumner High School in St. Louis. One could question why a scientist of such ability never held a position at a major research institution (Branson, 1955).

Ernest Everett Just (1883–1941) was a zoologist of exceptional ability. He completed about six scientific papers based on his research at the Marine Biological Laboratories at Woods Hole

(Massachusetts) from 1909 through 1916, before receiving his doctorate at Chicago, in 1917. Just spent most of his summers at Woods Hole, where he engaged in research, but in the early 1930s he left the laboratories at Woods Hole to do research abroad. In Germany he worked at the Kaiser Wilhelm Institut für Biologie (Berlin), in France at the Sorbonne and marine stations, and in Naples at the marine station. During this period the General Education Board, the Rosenwald Foundation, and the Carnegie Corporation supported his research. He published over sixty scientific papers during his career. In 1939 he published *The Biology of the Cell Surface* and also *Basic Methods for Experiments on Eggs of Marine Animals.* Yet he never received an appointment in a predominantly white institution. (He was a professor at Howard University from 1912 to 1941.) Some have suggested that the intensity of Just's immersion in scientific research was an attempt to escape his ascribed status of being black (Branson, 1955; Adams, 1964). F. R. Lillie, Just's major professor at Chicago, observed that "the numerous grants for research did not compensate for failure to receive an appointment in one of the large universities or research institutes. He felt this as a social stigma and hence unjust to a scientist of his recognized standing" (Lillie, 1942, p. 11).

Charles Drew (1904–1950) received an M.D. degree from McGill (Canada) and a D.Sc. degree from Columbia. He is most noted for his pioneering research in blood plasma. He discovered that plasma is often better for transfusions than whole blood. In 1941 he became director of the American Red Cross blood banks. At that time, ironically, blacks were not allowed to donate blood. Although protests by Drew and others resulted in a reversal of that policy, the blood of blacks was segregated and distributed solely to blacks. Drew spent most of his employed years at Howard University, a predominantly black institution. In 1950 he died as a result of injuries sustained in an auto accident near Burlington, North Carolina. He bled to death on the way to a "black" hospital after being denied admission to a "white" hospital (Branson, 1955; Adams, 1964; Lichello, 1968; Peters, 1970).

These brief histories may be isolated cases, but they raise a question about universalism in the social institution of science. Racial discrimination has been pervasive in United States society, and it is questionable whether science as a social institution can be generally free of that racism. While existing data are valuable for describing the geographical, undergraduate, and doctoral origins and the ultimate employment levels of black scientists, they do not permit a study of career patterns of black scientists. Without an analysis of career patterns, many doubts remain. If black Americans generally are discriminated against, why would not black scientists also be discriminated against?

Position of Blacks in Science

Bond (1972) estimated at the time of his study that the total number of living black academic doctorates (in all fields) was approximately 1,600 to 1,800 and that the annual production of black academic doctorates was roughly 160, or 1 percent of the total annual American output. Bryant (1970) reported similar results in a study conducted for the Ford Foundation; namely, that less than 1 percent of the Ph.D. candidates at the close of the 1967–68 academic year were black Americans. He concluded that, as of 1970, less than 1 percent of the total of America's earned doctorates were held by blacks and that, even if the number of blacks increased as much as 20 percent (for 1973), the figure would be less than 2 percent of the total of America's earned doctorates.

Various studies consistently report that black graduate students are highly represented in the field of education (see Bond, 1972; El Khawas and Kinzer, 1974; Brown and Stent, 1975; National Board on Graduate Education, 1976). In a study of the status of minorities in graduate education, the National Board on Graduate Education (1976, p. 4) reports an unusual phenomenon: that blacks tend to shift disciplines as they ascend the educational ladder. Of the 1973–74 black Ph.D.s who had an undergraduate major in the life sciences, about half (52 percent) continued in that field for doctoral study. In contrast, 80 percent of comparable whites earned a Ph.D. in the same field. The preferred choice of those black students who changed disciplines was education. In a survey of black graduate and professional school enrollments, Brown and Stent (1975) found that blacks accounted for only 3.4 percent of the 1970 graduate school enrollment in the United States. McCarthy and Wolfle (1975) point out that while native

American blacks tend to be overrepresented in education (in comparison to other disciplines), nonnatives tend to be more concentrated in the scientific fields.

In their chapter devoted to the topic of discrimination against women and minorities in American science, Cole and Cole (1973, pp. 152–154) remark that the extremely small number of blacks in science makes it difficult to study and generalize about black scientists. Nevertheless, they go on to point out that black scientists tend to receive their degrees from the less distinguished universities. Furthermore, if they have the doctorate, they are likely to hold appointments at historically black colleges and universities "far removed from the frontiers of scientific advance" (1973, p. 153).

The exact number of black science doctorates is, of course, unknown. Studies consistently reveal that blacks do not become scientists in a number proportionate to their percentage of the population. Although the black population in the United States is estimated to be about 12 or 13 percent, blacks are believed to constitute about 1 percent of America's total science doctorates. Richardson (1976) reports that the percentage of blacks holding the doctorate in the natural sciences has *never* exceeded 1 percent of America's science doctorates in any year. In fact, he asserts that the rate frequently has been only a small fraction of 1 percent. In his estimation, based on the best available information from 1973 to 1974, about 950 blacks earned the doctorate in the natural sciences (agricultural sciences, astronomy, biological sciences, chemistry, geological sciences, physics, and mathematics) from 1876 to 1976. According to Bond (1972), about 625 blacks received the Ph.D. in the sciences between 1920 and 1962. Jay (1975) estimated that the total number of black doctorates in the natural sciences as of 1975 did not exceed 1,200. The National Science Board (1975) reports that, as of 1974, 96 percent of all American scientists were white, 2 percent of the total were Asian, and roughly 1 percent were black.

To determine whether the low number of blacks pursuing the Ph.D. in psychology is due to regional or educational factors, Wispe and associates (1969) matched twenty-eight historically black colleges with twenty-eight predominantly white colleges on (1) public-private control, (2) geographical region, (3) number of faculty and students, (4) operating income, and (5) value of buildings. They found that black graduates of predominantly white institutions were more likely than black graduates of historically black colleges to pursue the Ph.D. The twenty-eight historically black colleges produced a total of 76 graduates who earned the Ph.D. during the period 1920–1966, while the twenty-eight white institutions produced 167 graduates who earned the Ph.D. during the same period. Black psychologists, like other black scientists, tend to earn only a nominal proportion of the Ph.D.s granted in their field. For example, between 1920 and 1966, blacks received less than 1 percent of the doctorates in psychology granted by the twenty-five universities giving the largest number of such degrees.

Blacks seem to have faired no better in sociology than in the other sciences. Conyers (1968) and Glenn and Weiner (1969) show that blacks do not contribute their proportional share to the pool of Ph.D. sociologists. Current estimates place the number of black holders of the Ph.D. between 1 percent and 2 percent.

Even after earning a Ph.D., blacks may have problems pursuing research. In their sample of 293 faculty holders of doctoral and master's degrees in political science (of whom 221 were white and 72 were black), Puryear, Woodard, and Gray (1977) report that whites were more likely to be the recipient of a research grant than their black peers. In fact, roughly half of their white subjects had at some point in their careers received a research grant, compared to only a fourth of the black faculty. They explain that white faculty are more likely than their black colleagues to begin receiving research grants at an earlier stage in their careers. Almost a third (33 percent) of the white faculty between the ages of 30 and 34 had received one or more research grants, but about 8 percent of the blacks in this age group had done so. White faculty consistently received more research awards at each age interval except for blacks over age 60. Moreover, the probability of receiving a grant is correlated with the size of an academic institution. For example, scientists in institutions with student populations of 20,000 or more tend to have a distinct advantage over those in smaller institutions, defined as those with populations of 1,000 or less. But Puryear and his colleagues (1977, p. 64) conclude that "the affiliation of a faculty member with a white institution, irrespective of quality, establishes a clear advantage for the faculty member whether he is black or white." What is not known, however, is whether these patterns exist for other science fields.

Origins of Black Doctorates

In a sample of 221 black scientists, Young and Young (1974) found that about seven out of ten were born and reared in southern states, while slightly more than one in ten came from the eastern states, and a nominal proportion had their geographical origins in the midwestern and border states. Jay (1971) reports similar results from his sample of 546 black natural science doctorates; 75 percent of his sample were born in the South, and the remaining 25 percent were born in the North and the West. This pattern in the geographical origins of scientists does not hold for nonblacks. For the period 1920–1961, Harmon and Soldz (1963) report that only 13 percent of all United States scientists had come from the South.

Jay (1971, p. 62) has offered some explanation for the large number of southerners among black natural science doctorates. First, he believes that southern blacks have traditionally been educated by other southern blacks, who have encouraged and motivated them to carry their training to the graduate level. Second, the existence of historically black colleges and universities as places of employment could have served as additional motivation to obtain graduate training. Third, southern blacks, having been exposed to black science doctorates in various administration positions, perceived that there were career opportunities and job security after completing the doctorate.

Wispe and others (1969) report on a study of 398 black American psychologists, of whom 166 had the Ph.D. degree. They found that the black psychologists in their study closely paralleled the distribution of the black population (for example, 52.8 percent of the black population reside in the southeastern region of the United States, and over half of their respondents resided there). In contrast, white psychologists tended to be overrepresented in the Northeast (36.1 percent to 24.4 percent) and underrepresented in the Southeast (15.4 percent to 23.7 percent).

The origins of doctorates may produce important differences in black scientists' careers. The differences in occupational choices between southern and northern blacks may be attributed to geographical origins. Pettigrew (1964, pp. 49–52) argues that, because southern blacks are exposed to few status ambiguities, they are more likely to accept or attempt to make adjustments to existing situations. In contrast, northern blacks, having been exposed to more status ambiguities, are more likely to desire to compete in an open market on the basis of equality.

Conclusion and Implications

The total number of black doctorates in all fields, including science, is relatively small (an estimated 1 percent of all of America's Ph.D.s). Furthermore, studies consistently reveal that blacks do not provide scientists in a number proportionate to their population size. While blacks account for roughly 12 percent of the nation's population, they are believed to constitute only 1 percent of America's science doctorates.

What we know about black scientists is limited. We do know, however, that black scientists tend to be male and southern-born. They received bachelor's degrees from historically black colleges or universities, completed their doctorates at a midwestern university, and now are employed in a historically black college or university. The number of black Ph.D.s employed in predominantly white colleges and universities, however, has shown a moderate increase since the late 1960s. In addition, there is some evidence that the salaries of black science Ph.D.s are beginning to reach parity with those of their white peers, and in some cases they actually earn more. Much of this progress is attributed to affirmative action programs.

What implication does discrimination have for the content and progress of science? In the "hard" and "soft" sciences, the demands for a black perspective are becoming increasingly noticeable (see Wilcox, 1971; Jones, 1972; Ladner, 1973; Young and Young, 1974; Smith, 1977). Smith (1973) has criticized his fellow black psychologists for applying to blacks the psychological theories developed on whites, causing us to label black youths incorrectly and thus lose talented black people who did not fit into the psychological conceptions based on the white experience.

Medical research has generally found cures and controls for most infectious diseases. But health institutions generally have not responded to the needs of the black community with similar results (Knowles and Prewitt, 1969). If the needs of the black community are to receive any significant response, given the lack of attention by the scientific community as historically constituted, the response may have to come primarily from blacks themselves or other concerned persons. In order to bring about any kind of change, a concerted effort will have to be put forth to increase the number of blacks in science.

A black perspective may bring some insight to various research problems that have previously been neglected. Evans (1977), for example, cites the case of sickle cell anemia research (a disease that primarily afflicts blacks). Funding for sickle cell research never exceeded $250 thousand until 1970. In 1971 Congress allocated $10 million for a concentrated research effort on sickle cell. According to Evans, this effort came about primarily as a result of a black HEW fellow, Colbert King, who happened to receive the assignment. Evans asserts that this situation illustrates the need for a more equitable representation on the advisory boards responsible for reviewing grant requests submitted to the National Institutes of Health. It also illustrates a general lack of concern for health problems affecting blacks. Melnick (1977) points out that minority groups and the poor are consistently exploited in biomedical tests and experimentation, primarily because they are powerless.

In 1973 black women had the fourteenth highest suicide rate in the world (an 80 percent increase in the past twenty-five years). According to Slater (1973a), there are only a few psychiatrists and sociologists specializing in suicide research; and, among these, only a small number are believed to be concerned with suicide among black people. This area also requires attention from black behavioral scientists, since white scientists seem to have little concern for it.

Similarly, a black perspective may serve to focus on the problem of hypertension, which kills an estimated 60,000 Americans annually (of whom about 13,500 are black). Hypertension is a major cause of death among black adults in this country. According to current estimates, about 25 percent of all black Americans and 15 percent of all white Americans age 18 years and older are afflicted with hypertension (Slater, 1973b).

Black males are believed to have the highest rate of prostatic cancer in the world. The number of black men dying of cancer is increasing. In fact, more than twice as many black men die of cancer today than they did thirty years ago. Black males have a higher incidence of cancer of the lung, stomach, esophagus, larnyx, prostate, tongue, small intestine, liver, pancreas, penis, and soft tissues. (White males, by comparison, have a higher incidence of cancer of the skin, brain, testis, bladder, and kidney.) It is believed that the cancer rates among black males are related to specific employment and dietary factors. A black perspective in medical sociology and biomedical science would have much to offer in the way of research in this problem (Slater, 1974).

Relatively scanty available data have prompted us to raise questions about the role performance of blacks in all areas of science. To get valid answers, we must look at the context of the role with different kinds of performance data, both at the predoctoral and postdoctoral levels. For example, studies have generally failed to consider how a scientist's first postdoctoral position is affected by variables such as predoctoral performance (measured in terms of publications), age at completion of the doctorate, and prestige of department. Further, we need to know more about how black scientists perform their postdoctoral roles (measured in terms of publications and citations to their work). Only through such study will we be able to assess accurately the role performance of black scientists and the presence or absence of racial discrimination in science.

The type of research suggested would provide some basis for dealing with the problem of how to encourage more blacks to go into science. For example, if black college and high school students have the results of a comprehensive systematic comparative study of black and white scientists available to them, it could have a positive influence in attracting more blacks into science. Science might be seen as a channel of upward mobility not subject to structural blockages that other avenues of mobility often present to blacks (Blalock, 1967; Pearson, 1977). If, in contrast, black scientists are not treated in a universalistic fashion, then the mechanisms of particularism and the outcomes of particularistic evaluations of role performance, and the subleties of such particularism, may be identified. If discriminatory mechanisms are present and can be identified, countervailing forces may be instituted in science to remedy such particularistic treatment of blacks in the future. A large-scale research project that began in the fall of 1977 should eventually provide answers to many of the questions discussed in this chapter.

References

Adams, R. L. *Great Negroes: Past and Present.* (2nd ed.) Chicago: Afro-American Publishing Co., 1964.

Blalock, H. M., Jr. *Toward a Theory of Minority-Group Relations.* New York: Wiley, 1967.

Bond, H. M. *Black American Scholars: A Study of Their Beginnings.* Detroit: Balamp, 1972.

Branson, H. R. "The Negro and Scientific Research." *Negro History Bulletin,* 1952, *15* (7), 131–137.

Branson, H. R. "The Negro Scientist." In J. H. Taylor, C. Dillard, and N. K. Proctor (Eds.), *The Negro in Science.* Baltimore, Md.: Morgan State College Press, 1955.

Brown, F., and Stent, M. D. "Black Graduate and Professional School Enrollment: A Struggle for Quality." *Journal of Black Studies,* 1975, *6* (1), 23–33.

Bryant, J. W. *A Survey of Black American Doctorates.* New York: Ford Foundation, 1970.

Clemente, F. "Race and Research Productivity." *Journal of Black Studies,* 1974, *5* (2), 157–166.

Cole, J. R., and Cole, S. *Social Stratification in Science.* Chicago: University of Chicago Press, 1973.

Cole, S., and Cole, J. R. "Scientific Output and Recognition: A Study in the Operation of the Reward System in Science." *American Sociological Review,* 1967, *32,* 377–390.

Cole, S., and Cole, J. R. "Visibility and the Structural Bases of Awareness of Scientific Research." *American Sociological Review,* 1968, *33,* 397–413.

Conyers, J. E. "Negro Doctorates in Sociology: A Social Portrait." *Phylon,* 1968, *29* (3), 209–223.

Crane, D. "Scientists at Major and Minor Universities: A Study of Productivity and Recognition." *American Sociological Review,* 1965, *30,* 699–714.

El Khawas, E. H., and Kinzer, J. L. *Enrollment of Minority Graduate Students at Ph.D. Granting Institutions.* Washington, D.C.: American Council on Education, 1974.

Evans, T. E. "An Informed Constituency with a Representative Bureaucracy: Health Policy and Black People." In V. L. Melnick and F. D. Hamilton (Eds.), *Minorities in Science: The Challenge for Change in Biomedicine.* New York: Plenum Press, 1977.

Freeman, R. B. *Black Elite: The New Market for Highly Educated Black Americans.* New York: McGraw-Hill, 1976a.

Freeman, R. B. *The Overeducated American.* New York: Academic Press, 1976b.

Gaston, J. *The Reward System in British and American Science.* New York: Wiley-Interscience, 1978.

Glenn, N. D., and Weiner, D. "Some Trends in the Social Origins of American Sociologists." *American Psychologist,* 1969, *4,* 291–302.

Hargens, L., and Hagstrom, W. "Sponsored and Contest Mobility of American Academic Scientists." *Sociology of Education,* 1967, *40,* 24–38.

Harmon, L. R., and Soldz, H. *Doctorate Production in the United States Universities 1920–1962.* Washington, D.C.: National Research Council, 1963.

Jay, J. M. *Negroes in Science: Natural Science Doctorates, 1876–1969.* Detroit: Balamp, 1971.

Jay, J. M. "Letter." *Science,* 1975, *190,* 834, 836.

Jones, R. L. (Ed.). *Black Psychology.* New York: Harper & Row, 1972.

Knowles, L. L., and Prewitt, K. *Institutional Racism.* Englewood Cliffs, N.J.: Prentice-Hall, 1969.

Ladner, J. A. (Ed.). *The Death of White Sociology.* New York: Random House, 1973.

Lichello, R. *Pioneer in Blood Plasma: Dr. Charles Richard Drew.* New York: Julian Messner, 1968.

Lillie, F. R. "Obituary." *Science,* 1942, *95,* 10–11.

McCarthy, J. L., and Wolfe, D. "Doctorates Granted to Women and Mintority Group Members." *Science,* 1975, *189,* (12), 856–859.

Melnick, V. L., "Public Policy for Minority Self-Actualization: Present Realities and Future Possibilities." In V. L. Melnick and F. D. Hamilton (Eds.), *Minorities in Science: The Challenge for Change in Biomedicine.* New York: Plenum Press, 1977.

Merton, R. K. "Science and the Democratic Social Order." In *Social Theory and Social Structure.* New York: Free Press, 1957.

Merton, R. K. *The Sociology of Science: Theoretical and Empirical Investigations.* Chicago: University of Chicago Press, 1973.

Mommsen, K. G. "Black Ph.D.s in the Academic Market Place: Supply, Demand, and Price." *Journal of Higher Education,* 1974, *45,* (4), 253–267.

National Board on Graduate Education. *Minority Group Participation in Graduate Education.* Washington, D.C.: National Academy of Science, 1976.

National Science Board. *Science Indicators 1974.* Washington, D.C.: U.S. Government Printing Office, 1975.

Pearson, W., Jr. "Blacks in Science: Why Aren't There More?" Paper presented at the Mid-South Sociological Association, Monroe, La., Nov. 1977.

Peters, M. *The Ebony Book of Black Achievement.* Chicago: Johnson Publishing Co., 1970.

Pettigrew, T. F. *A Profile of the Negro American.* New York: Van Nostrand, 1964.

Puryear, P. L., Woodard, M. C., and Gray, V. "The Comparative Status of Black and White Political Scientists." In M. C. Woodard (Ed.), *Blacks and Political Science.* Washington, D.C.: American Political Science Association, 1977.

Richardson, F. C. "A Quarter Century of Black Experience in the Natural Sciences, 1950–1974." *Negro Educational Review,* 1976, *27* (2), 135–154.

Slater, J. "Suicide: A Growing Menace to Black Women." *Ebony,* 1973a, *28,* 152–160.

Slater, J. "Hypertension: Biggest Killer of Blacks." *Ebony,* 1973b, *28,* 74–82.

Slater, J. "The Rise of Cancer in Black Men." *Ebony,* 1974, *29,* 92–100.

Smith, S. L. (Ed.). *Black Political Scientists and Black Survival.* Detroit: Balamp, 1977.

Smith, W. D. "Which Way Black Psychologists: Tradition, Modification, or Verification-Innovation?" *Journal of Black Studies,* 1973, *4,* 3–7.

Taylor, J. H., Dillard, C., and Proctor, N. K. (Eds.). *The Negro in Science.* Baltimore, Md.: Morgan State College Press, 1955.

Wilcox, R. C. (Ed.). *The Psychological Consequences of Being a Black American: A Sourcebook of Research by Black Psychologists.* New York: Wiley, 1971.

Wispe, L., and others. "The Negro Psychologist in America." *American Psychologist,* 1969, *24,* 142–150.

Young, H. A., and Young, B. H. *Scientists in the Black Perspective.* Louisville, Ky.: Lincoln Foundation, 1974.

3

Donald R. Ploch

Research Funding for Sociology in the National Science Foundation

In recent years support for scientific work, especially federal support, has been examined and challenged on grounds of "fairness" or universality and specificity of procedures. As a result, most government agencies have been forced to examine distributions of funds, as well as the procedures for decisions that lead to these distributions. Many of the questions addressed and data gathered are important for the sociology of science. I will discuss some instances of "applied" sociology of science in the sociology program of the National Science Foundation (NSF). My knowledge is that of a former insider; for five years, ending in September of 1977, I was program director for sociology, at the foundation. These are personal reminiscences, not based on official files. They are my own views and not those of NSF. Moreover, since I no longer have access to the data or analyses discussed here, I can certify the figures only in general terms.

The Proposal Review Process

The proposals received by the National Science Foundation are generally unsolicited. They may simply be letters of inquiry or visits with the staff to determine whether "NSF is funding . . . ?" A "No" answer is reserved for projects that clearly belong to another government agency. All other advice about the probability of funding, given the state of the field or the quality of the proposals, is personal. As a result, each program director develops a personal style. The effectiveness of this style as a screen, to admit some proposals and push others out, has not been studied. (Such a study would require a substantial investment in field observation resources.) As a personal style, I tried to screen out as few proposals as possible, on the assumption that other elements of the review would flag deficient proposals.

All proposals are sent to a central office for identification, construction of a computer file (used to monitor their path through NSF), and assignment to a program. Persons sending proposals may request assignment to specific programs or may request reassignment. In many cases program directors will ask colleagues to read a proposal with a view to transferring it to another program or considering joint review and funding. Almost all proposals are assigned for review. During my tenure, less than five proposals were refused by the sociology program and all other programs which were asked to accept them. Once a proposal has been accepted for review, it is eligible for funding on scientific grounds; that is, it cannot be rejected as inappropriate for the program.

In sociology (and the social sciences generally), there is an advisory panel that meets three times a year to review proposals. During most of my tenure, there were five panelists, but the number was expanded to seven in 1977. The panelists generally serve two or three one-year terms, with terms overlapping to maintain continuity. Panelists are nominated by the program director after consultation with present panelists and active researchers. Appointments to the panel are the responsibility of the director of NSF, but for sociology that responsibility is usually delegated to the assistant director for biological, behavioral, and social sciences. Panelists are chosen to represent wide expertise across the discipline as well as to achieve a politically acceptable balance of ascribed characteristics, geographical distribution, and the like. The attempt to achieve such balance has not destroyed the effectiveness of the panel, since there are ample numbers of highly competent persons with desirable characteristics.

Each proposal also is reviewed by several ad hoc reviewers (usually three to five) who are expert in the field. They are generally given three weeks to respond, and about 60 percent of them do. Here again an effort is made to maintain a balance of ascribed characteristics—to select reviewers from four-year colleges as well as graduate departments, nonacademics as well as academics, from various geographical areas—and to avoid using the same reviewers over and over again. One of the responsibilities of the program director is to stay current with the field, so that he can incorporate new persons into the pool of reviewers.

Both ad hoc reviewers and panelists are advisers to NSF. The program director's recommendations—to fund, to fund partially, or to decline a proposal—are based on the advice given but are not bound by it. When the program director does not follow the advice of reviewers, however, he must provide a detailed justification for his decision, which is reviewed at other levels of NSF.

All recommendations by the program director are reviewed by the division director. If the division director decides to decline a proposal, his decision—as well as the recommendation of the program director—can be appealed through at least two levels within NSF. All award recommendations are reviewed by a special board. Either the division director or the special board or both may recommend, or insist on, changes before sending the recommendation to the assistant director. Once the assistant director approves, the scientific review is complete, though other offices (for instance, the financial, legal offices) conduct their own reviews. On completion of all reviews, the award is made when the senior grant official, representing the director of NSF, signs the proposal.

Grants which are larger than a set figure (about $500,000 per year or $2 million over all) or which set precedents in policy are reviewed by the board of directors of the foundation. They can and do recommend changes or additional review. Sociology has had only one project taken to the board. That project, to fund the production of public-use samples (1 percent) from the 1940 and 1950 censuses, was approved and is being carried out.

Program directors are hired to civil service–excepted positions because NSF wants active professionals in these positions. The expectation is that they will be temporary (one or two years), returning to traditional research activities. As professionals, program directors are given wide latitude to develop programs to take advantage of advances in the field. They are considered professional scientists and are given a relatively free hand in their interactions with researchers.

"Applied" Sociology of Science

All "applied" sociology of science on NSF funding is done by employees of NSF, either employees in the usual sense or persons hired for special purposes. Grantees—such as the Cole brothers (1977), who were funded by the National Academy of Science with funds from NSF—retain considerable independence to design research and publish results. Still, certain elements of the files, essential for research, are unavailable except to employees.

NSF publishes lists of awards by school and lists of reviewers by science area. It does not publish declines or reveal reviewers' recommendations for specific proposals. As a result, an external cannot evaluate funding decisions or distributions unless he judges the distribution of awards against some theoretical or "politically" desirable distribution. Empirically, one will find that funding is concentrated. That is what one would expect if he assumes a concentration of research excellence. But it is contrary to any expectation based on an assumption of uniform or random distributions of talent. Publicly available data are not sufficient to evaluate many such crucial questions of scientific import or public policy. I was struck by the fairness of the decision process as I knew it and by the concentration of funds as well as the criticisms of the program. The problem was to evaluate a set of decisions to determine characteristics associated with awards and declines.

During the summer of 1976, Carl Bennett Backman, then a graduate student at Cornell, was hired to help evaluate the sociology program. We analyzed all proposals submitted in fiscal 1975 (July 1, 1974, to June 30, 1975) that were subjected to review. This universe consisted of all new regular research proposals and all continuations for a year or more. It excluded all committed renewals and supplementary awards. Committed renewals, for a period not to exceed five years, are monitored by the program director, who recommends the approval of annual budget requests if the project is progressing satisfactorily. Supplementary requests are generally for small sums and short periods of time and are intended to bring a project to a satisfactory conclusion. Program directors usually use their own discretion in recommending these awards. Normally, only one supplementary award is made to any grant.

Proposals in the universe accounted for 90 to 95 percent of the distributed funds except for committed renewals. Since these renewals accounted for 20 to 25 percent of the funds, the universe examined accounted for 65 to 75 percent of the total budget, and about the same percent of program decisions. The remainder consisted of committed renewals (10 percent), dissertations (10 percent), travel awards (5–10 percent), and supplements (5–10 percent). The universe consisted of about 106 proposals. We tried to evaluate the award-decline decision against research area, size of request, school of Ph.D. degree, professional age, rank, chronological age, and publications of articles and books.

Eighty percent of the proposals were declined, so the best prediction was declination. None of the indicators, singly or in combination, appreciably improved this prediction. Two indicators were significantly related at the 5 percent level; each reduced error about 1 percent, so the final predictive ability was about 82 percent. These indicators were publication of articles in the *American Journal of Sociology (AJS)* and the *American Sociological Review (ASR)* within the past five years and prestige of current institution.

Professors who had not published in *AJS* or *ASR* in the past five years had 10 percent of their requests funded. Those who had published one or two articles in these journals in the past five years had 20 percent of their proposals funded. Proposers with three or more articles had 50 percent of their proposals funded. The differences are large compared to the predictive ability of the variable, because most proposers have not published and very few have three or more articles.

Many related variables had no significant effect, and at least one effect was quite contrary to expectations. Publication of books had a curvilinear effect, with the lowest probability of funding associated with the publication of one or two books. Proportions funded were higher for those with no books and for those with three or more. Total bibliography did not predict, nor did publication in *AJS* and *ASR* for any period except the last five years.

I do not remember detailed figures for the effect of prestige of current institution. But the impact on decisions, though statistically significant, is substantively small. My interpretation of these findings is that award decisions were made in favor of those proposers (for proposals) who directed their remarks to the central core of sociology as defined by major journals. On the whole, the research was narrowly specified so that it could be accomplished in a few years. It was also well related to the general structure of the field. It constituted work whose thrust was more acceptable to *AJS* or *ASR* than to specialty journals. It appealed to the community of sociologists rather than only to the scholars in a particular specialty area. It was sociology rather than social statistics, ethnomethodology, demography, or delinquency.

The null findings are especially important. Older investigators are not favored, nor are large budget requests, nor full professors, nor prestige institutions. Substantively, one is left with the fact that award-decline decisions are practically unrelated to the characteristics of proposers or their institutions.

Further analysis showed that we could predict awards and declines separately but that these predictions were the same. Thus, we were predicting proposal submission. Persons at prestigious institutions, well published, older, and higher in rank submit more proposals and are funded more and declined more than others. External analysts are correct in their analysis of awards, but without access to decline decisions, those analysts cannot know that theirs is really an analysis of proposals. At least in the sociology program, the award-decline decision for proposals submitted in fiscal 1975 appears to be fair, in that it is not related to extraneous characteristics of institutions or individuals.

If the decision is trichotomized into categories of award, decline for lack of funds, and decline for other reasons, the substance of the results does not change, though some interesting findings arise. Younger proposers and those from less prestigious institutions tend to be given awards, or they are declined for lack of funds. Older proposers and those from prestigious institutions tend to be granted funds or declined for other reasons. The older proposers are clearly in or out of the running; the younger proposers fall into a gray area. Proposals from these applicants are acceptable scientifically, but the program budget would have to expand by a factor of two or three to be able to fund them at any reasonable level.

We did not evaluate individual reviews because NSF was (and probably still is) reluctant to develop files linking reviewer recommendations to proposals. The anonymity of individual reviews is a sacred trust. Because our research had to be completed in three months, there was no time to secure clearance for such a file even on a pilot scale. We set up coding forms for the data, but they were never used. In addition, we did not code the aggregate reviewer score of a proposal, primarily because I felt it was a senseless enterprise. Reviewers were allowed to give an overall rating of a proposal on a five-point scale from excellent to poor. The majority of programs in NSF code these 1 to 5 and produce averages, sometimes to two and three decimal places. The scales are ordinal at best and have never been tested for interrater reliability. Thus, our producing an average score seemed pointless and in violation of accepted principles. In hindsight, this decision was a mistake because other evidence suggests that the average score predicts well and does address issues of program director authority.

I did some other research that bears on this point. In the fall of 1975, the panel was entirely new. The panel members wanted to rank all proposals to clarify their own work and guide my decisions. I resisted this suggestion, and we compromised on an experiment. Each panelist ranked no more than twelve proposals in the order in which they would be funded. These lists were given to my assistant to hold until I had made funding decisions based on all reviews and panel discussion. After my decisions were made, we merged the lists by scoring one point for a proposal that appeared on a list and by averaging ranks after standardization for lists of different lengths.

Of forty-two proposals funded, only one was not recommended by any panelist. In funding that proposal, I negotiated a reduction of time, from three years to one, and the budget, which was reduced by a factor of ten. The first four proposals on the master list, and six of the first eight proposals, were funded. One of those not funded was withdrawn because funds were received from other sources. All funded proposals, except the one discussed earlier, ranked in the top fifteen. At the next meeting the panelists agreed that funding the low-ranking proposal in the manner I did was appropriate. Funding decisions followed their rankings closely enough that the experiment was never repeated.

Cole, Rubin, and Cole (1977) show that 78 percent of the variation in award-decline decisions in ten programs at NSF is explained by the average score. There is no other variable with comparable power. The conclusion from these two studies is that if the program director exercises despotic power, then he does so most of the time with the advice and consent of reviewers. Put another way, the bulk of the award-decline variance is accounted for by characteristics of the proposal.

The selection of reviewers and panelists and their individual evaluations are not well studied. (See Cole, Rubin, and Cole, 1977, for the best work to date.) In the mid 1970s, there was much talk of random selection, blind reviews (in NSF all proposals are fully identified to reviewers), and other changes in the review process. The controversy appears to have died down, but it is not forgotten. One could easily compile a list of persons who were asked to review and their responses. Such information is generally available in individual proposal files. The more difficult task is to observe or interview a program director to determine the rationale used to select reviewers. Most could elaborate an algorithm for selection but probably could not indicate how closely it is followed. Close observation might show substantial variation from the algorithm from time to time. We need descriptive data from superb naturalists before we can begin effective evaluation.

Interrater reliability is also a major problem. The myth is that interrater reliability goes down, or variance in overall ratings goes up, as one progresses from natural sciences through life sciences to social sciences. There is enough truth to keep the myth alive and circulating, although the fields are not as far apart as the myth holds. Cole, Rubin, and Cole (1977) report that the myth does not hold for their study. This is an area for future research.

That funds are unevenly distributed is a well-known fact.

Some researchers get more than others. The sources of this unevenness, be they excellence or ability or luck or power, cannot be examined with data in the public domain. NSF and other agencies would be well advised to publish information on declines as well as awards. Current arguments against this practice are weak. NSF does not want to harm people by publishing their failures. But, given the submission procedures of most institutions, this information is generally available at the local level. To make it available nationally might lead to abuses (such as denial of tenure); but these are likely to be few, since many other factors are taken into account in any decision. (Tenure, for example, is not likely to be judged solely on proposal success.) Though the research is proprietary, little seems lost by publishing titles, dollars requested, institution, and proposer.

If declines were published, external analysts could do a better job of evaluating the basis of decisions. Such a procedure would certainly alter the environment in which NSF operates, but there is no evidence that its effectiveness would be diminished.

Conclusions

Although effective management of NSF and sociological studies of science will often use the same data in similar analyses, information should be made public because scientific studies will have aims not shared by management, which often must choose between analyzing its work and doing its work. Regardless of the hue and cry about "fairness," there would be a louder and more justified cry if NSF ignored its work in order to analyze how the work was done. In any event, my analyses show that in sociology awards are drawn from the same universe as declines. The decision is "fair," at least with respect to proposals submitted. Concentration of awards or correlations of awards and other characteristics are functions of proposal submissions.

References

Cole, S., Rubin, L., and Cole, J. R. "Peer Review and the Support of Science." *Scientific American,* Oct. 1977, *237,* 34–41.

Part 2

Problem Choice

How do scientists choose their research problems? Why are some areas of research apparently neglected and other areas worked on by many researchers? How should scientists choose research problems? What considerations should determine the theoretical perspective used by scientists? Answers to these questions will go a long way toward explaining the social influences on scientific change, and these kinds of questions concern both philosophers of science and sociologists of science.

If—among its other aims—the sociology of science uses sociological concepts and theories to try to explain that part of science amenable to social rather than cognitive investigation, then the chapters included here provide the foundation for proceeding with the task.

Harriet Zuckerman poses questions about the sociology of science's possible benefit to the philosophy of science. She examines the considerations involved in theoretical and problem choices in actual scientific research and offers novel ideas on the way that theory can inhibit important empirical inquiry. It may be that the philosophy of science tries to deal with someone's creation of an ideal world of science (although there is no consensus on such a world), whereas the sociology of science tries to deal with the world as it is. As a sociologist of science, Zuckerman may succeed in creating the kind of interaction between sociologists and philosophers that will help to find answers to critical questions in each discipline.

Thomas F. Gieryn looks at the process whereby scientists change problems during their careers. He also considers the types of problems that scientists select and the social conditions that influence the potential set of choices. Science has been considered, far too much, as an activity whereby scientists are free to choose their topics with minimum constraints. But do scientists, in fact, have—or need—total freedom in research? Few would disagree that significant research requires freedom of inquiry, but the question remains: Under what conditions, and at what stage of research, is autonomy in science imperative? Gieryn's chapter provides a systematic and comprehensive overview of the typologies that are theoretically possible during the careers of researchers. His discussion of the alternative patterns and the consequences for scientists' careers of the choices they make, and for scientific knowledge, will assist other researchers in designing comparative studies on additional disciplines and over longer historical periods.

4

Harriet Zuckerman

Theory Choice and Problem Choice in Science

Research in the sociology of science these days cannot easily ignore related perspectives in the philosophy of science, just as research in that field has grudgingly begun to include sociological perspectives. (For a recent discussion of actual and recommended connections between these fields, see Barber, 1977.) These conjoint perspectives are particularly evident in efforts to describe and account for cognitive change in science. It may be timely, then, to review current sociological and philosophical work bearing upon cognitive change so as to identify their potential convergences while recognizing their inevitable differences.

Note: Support for this study was provided by a National Science Foundation grant (SOC 77-172373) to the Columbia University Program in the Sociology of Science. An early draft was read at the 72nd annual meeting of the American Sociological Association in September 1977 and a later one at the colloquium of the department of the history and philosophy of science at Princeton University in April 1978. I am indebted to Donald Campbell, Charles Gillispie, Thomas Kuhn, Joshua Lederberg, Stephen Toulmin and especially to Robert Merton, for their critical comments.

Practically all sociological work on cognitive change in science focuses on one or more of three sets of questions: (1) theory choice and evaluation (how scientists choose among competing theoretical orientations and how they appraise proposed contributions to a research field); (2) problem choice (how scientists select problem areas in which to work and specific problems within them) and (3) resulting cognitive groups (how shared theory choices make for the development of "theory groups" and how shared problem choices contribute to the formation of invisible colleges and specialties).

Because philosophers of science, with the important exceptions of Thomas Kuhn (1962, 1974, 1977) and Stephen Toulmin (1972), have far more often focused on theory choice and problem choice than on their related social formations, these types of recurrent decisions provide apposite materials for comparing philosophical and sociological perspectives on scientific change.

Theory Choice

In dealing with questions of theory choice, contemporary philosophers of science, at times it seems unwittingly, tend to oscillate between normative and descriptive accounts: how scientists should choose among competing theories and how they actually do choose.

Among normative epistemological accounts, Karl Popper's doctrine of "falsifiability as the touchstone of scientific rationality," with its antecedents in the writings of Norman Campbell and William Whewell, has become a major focus for debate. Popper makes falsifiability a criterion for marking off empirical sciences from nonscience (particularly what he describes as pseudoscience—for example, psychoanalysis and Marxism); moreover, the capacity of a scientific theory to withstand attempted refutations is, in his view, the prime determinant of its probability of survival. As Popper observes, "By passing such tests of [relative severity], a theory may 'prove its mettle'—its 'fitness to survive'" (1974, Vol. 1, pp. 1, 82). In the Popperian view, the theory that survives the most demanding test will be selected, although the notion of differing severity of tests itself requires clarification.

For the sociology of science, it is of direct interest that Popper (1959, 1962, 1974) in effect adopts a universalistic criterion for both recommended and actual choices among contesting theories and other scientific contributions. His criterion for assessing the relative strength of contesting theories is their success in surviving attempted refutation in the form of empirical test and "critical thought." This is universalistic in the strict sense that "the acceptance or rejection of claims entering the lists of science is not to depend on the personal or social attributes of their protagonists: their race, nationality, religion, class, and personal qualities are, as such, irrelevant" (Merton, [1942] 1973b, p. 270). Popper's falsifiability doctrine thus links up tacitly but strongly with sociological investigations that have focused on the extent to which universalistic or particularistic criteria are, in fact, used in accepting or rejecting proposed contributions to science.

Aspects of Popperian doctrine have come under heavy fire from many sources, not least from students and colleagues such as Imre Lakatos (1974) and Joseph Agassi (1977) and perhaps most forcefully from Adolf Grünbaum (1976a, 1976b). I need not go into the details of this controversy. From the perspective of the sociology of science, however, the controversy *itself* provides raw materials for a case study of how philosophers of science choose among rival theories. It raises the question of whether the processes involved in their decisions differ from processes of theory choice among scientists. Empirical investigation of the matter would connect the problem involved in actual theory choice in science and philosophy with the problem of demarcation, of drawing boundaries between them. Current debates in epistemology can thus provide data for sociological research on theory choice. Indeed, in an interchange with Popper, Kuhn (1974, Vol. 2, p. 802) in effect also proposes the beginnings of a hypothesis for such research:

> The tradition of critical discussion [which Sir Karl] represents [as] the only practicable way of expanding our knowledge" . . . does not at all resemble science. Rather it is the tradition of claims, counterclaims, and debates over fundamentals which

> ... have characterized philosophy and much of social science.... In a sense, to turn Sir Karl's view on its head, it is precisely the abandonment of critical discourse that marks the transition to a science. Once a field has made that transition, critical discourse recurs only at moments of crisis when the bases of the field are again in jeopardy....
>
> Only when they must choose between competing theories[1] do scientists behave like philosophers. That, I think, is why Sir Karl's brilliant description of the reasons for the choice between metaphysical systems so closely resembles my description of the reasons for choosing between scientific theories.

Kuhn refers, of course, to his widely familiar reconstruction of how scientists choose between competing paradigms and between theories within them—with its central notion that such choices are constrained by prior cognitive commitments that lead adherents of differing paradigms to "talk past each other."[2] Systematic investigation of how scientists and philosophers actually choose among theories now seems in order.

Kuhn's conception, in turn, has been subjected to strong criticism (for instance, by Lakatos, 1971, 1974), particularly for its relativistic implications. But it is not so much processes of choice between competing theories and hypotheses as the choice between competing paradigms (despite the controversial assumption of incommensurability between them) that has elicited much discussion and even some research in the sociology of science.

Empirical sociological research bearing on theory choice has focused on two quite different sets of issues: (1) how the evaluation system operates to appraise and sort out theories and other scientific contributions; (2) how theories and theoretical orientations emerge, attract new recruits, and eventually become displaced by new theories.

The evaluation system in science is composed largely of institutional arrangements for the exercise of organized skepticism (Merton, 1973e, passim)—arrangements such as the referee process in scientific journals; peer review of proposals for research

theory was never decisively refuted but was rejected on the assumption that the J phenomenon would eventually be explained on routine grounds and did not require adoption of as "revolutionary" a theory as Barkla's. Apparently the J phenomenon was troublesome for existing theory but not troublesome enough to elicit sustained research; and other related observations being made at the time did not call existing theory into enough question to justify radical change.

As throughout this truncated review, pertinent studies must be bypassed, such as those on resistance to scientific discovery (Barber, 1961; Merton, [1963] 1973a, chap. 17; Cole, 1970).[3] But empirical studies of theory choice—including patterns of acceptance, resistance, or rejection—must develop adequate, not merely feasible, indicators of explanatory concepts such as "heuristic potentials" and "nonrational resistance" to theories. To be obvious and blunt about it, no good can come of empirical studies that aim to use current concepts and ideas in the philosophy of science if great gaps remain between those concepts and the indicators actually employed in the research. Recent studies of the complex relations between theory and empirical investigation in science (see, for instance, Edge and Mulkay, 1976; Sullivan, White, and Barboni, 1977; Sullivan, Barboni, and White, 1978) testify to the renewed awareness that this methodological problem must be placed high on the research agenda of the sociology of science. Until empirical studies move beyond their present primitive state in the devising of adequate indicators, we cannot draw research-based conclusions about the aptness of descriptive, much less normative, philosophies of theory choice in science. Turning from studies of *theory* choice, we want to consider how sociologists and philosophers account for *problem* choice in science.

Problem Choice

Scientists occasionally report that finding significant research problems is far more difficult than finding their answers. (For the import of such observations, see Merton, 1959, pp. ix-x.) But if scientists largely agree on this, they evidently can differ in identifying what are and are not scientifically consequential and soluble problems. In trying to identify ingredients for a sociological definition of a scientific problem, therefore, we take as a beginning what scientists say they are working on or could be working on.[4] Such a definition, while wanting in many respects, meets one important criterion; that is, a problem is scientific only if it is considered amenable to investigation. (On science as "the art of the soluble," see Medawar, 1967, and Hagstrom, 1965, pp. 277–278, for cases in point.)

In his study of science in seventeenth-century England, Merton ([1938] 1970, chaps. 7-10, appendix) took as one of his principal problems the identification of cognitive and extrascientific influences upon problem choice in science. More recently, sociologists have begun to adopt the self-exemplifying stance that problem choice must be a central problem in studies of scientific development (Zuckerman, 1974; Edge and Mulkay, 1976; Weinstein, 1976; Edge, 1977; Gieryn, 1977; Sullivan, Barboni and White, 1978; see also the chapter by Gieryn in this volume). Current sociological investigations of problem choice relate to three sets of questions: (1) In actual practice rather than methodological precept, how do some problems come to be defined as interesting or even as basic and others as uninteresting? What leads to a question's being defined as sensible or absurd? (2) Apart from occasionally originating new problems (see Merton, 1959, pp. xi–xii), how do scientists choose from the pool of identified problems? (3) How do theoretical contributions defined as pathbreaking in a discipline or area of research affect subsequent problem choices?

Though put in sociological rather than philosophical terms, these queries echo epistemological issues raised by philosophers of science—issues such as appropriate connections between paradigms or research programs and problem choice in science or appropriate interactions between research findings and theoretical formulations.

In spite of widespread skepticism about "methodological precepts," recent case studies of problem choice often find in actual practice what methodological dogma has long maintained should be the case: that scientists define some problems as pertinent, and others as uninteresting or even illegitimate, primarily on the basis of theoretical commitments and other assumption structures. Empirical case studies find that theory and its associated concepts and labels can preempt research attention in the following ways: (1) by defining certain observations as irrelevant or incorrect and therefore not worth following up because they are inconsistent with prevailing theoretical views; (2) by specifying certain investigations as unfeasible, not possible to carry through given the state of the experimental art, and thereby foreclosing work along those lines; (3) by fixing scientists' perceptions of aspects of nature through reification of concepts and labels, thereby precluding the possibility that certain phenomena might be seen as other than what those concepts and labels imply; (4) by constraining definitions of what is and is not problematic in the first place; and (5) by diverting scientists' attention from some problems because theory has sharply focused attention on other problems.

Theoretical preemption of the first sort seems to have been responsible for physicists' lack of interest in experimental evidence (published in 1928 by R. Cox, C. McIlwraith, and B. Kurrelmeyer), which we now know called into question the important and long-standing law of parity conservation. It was not until 1956–57, when T. D. Lee and C. N. Yang questioned the validity of parity conservation in weak interactions on theoretical grounds (for reasons unrelated to the Cox experiment) and when their work was experimentally supported by C. S. Wu and her collaborators, that the early experiment was dredged up and its significance newly identified. (See Yang, 1958; Bernstein, 1962.)

Another case of theoretical preemption, this time involving the foreclosure of empirical work, is found in Edge and Mulkay's (1976) study of the prehistory of radio astronomy. They point out (pp. 9–11) that four unsuccessful attempts to observe solar radio emissions were made between 1890 and 1901.[5] Such emissions should have been detectable, since Hertz had observed radio waves in 1887, and Maxwell had predicted their presence some fourteen years before. Yet, after those four unsuccessful attempts, there is no record of further work for thirty years. Are these experimental failures enough to explain the thirty-year-long hiatus in research? Perhaps, but there may have been another compelling reason. Following the physicist Westerhout, Edge and Mulkay (1976, p. 10) suggest that "widespread acceptance of Planck's theory [of 1902] discouraged research into extraterrestrial radio waves. As Westerhout (1972, pp. 211–212) explains: "If Planck's theory were correct, it followed that the radiation from the sun should be blackbody radiation, and the radio emission from a 6000°K blackbody was, of course, undetectable at that time. Similarly, radiation from the stars should be undetectable and one could calculate that the planets and the Milky Way could not possibly give any detectable signals either." In this instance, then, theory served to foreclose further experiments on the problem; for Planck's theory seemed to explain the repeated experimental failures and provided grounds for supposing that experiments were unfeasible.

When, after the hiatus of three decades, experimental work on the problem was resumed, it was not because scientists had raised themselves above the theoretical preconceptions induced by the Planck doctrine to engage in active search for these radio emissions. Rather, the renewal of interest in the problem was brought about by a serendipitous discovery. Radio emissions from the Milky Way were first recorded in 1931 by Karl Jansky of the Bell Laboratories while he was engaged in the eminently practical program of searching for the source of static in transoceanic telephone communications (Jansky, 1932, p. 1921). It is not at all evident that Jansky, an engineer, had to counter the preemptive assumptions of Planck's theory in order to get on with his work. At any rate, he did not call attention to the Planck doctrine in his papers; and, although he plainly realized that his observation was of importance (Friis, 1965), he expressed no sense of surprise in the significant sense of regarding the phenomenon as theoretically anomalous. (On serendipity as involving an "unanticipated, anomalous, and strategic datum," see Merton, 1968c, pp. 157–162.)

As this case suggests, accidental discoveries can counteract the results of theoretical preemption. They can bring into question, at least temporarily, theoretical commitments and derivative assumptions that have introduced theoretical blind spots and foreclosed certain lines of inquiry as not meriting empirical investigation.

Although Jansky's findings received considerable public attention (they made front-page news in the *New York Times* on May 5, 1933, and his "galactic hiss" was even broadcast on the radio), just one physicist and two astronomers (Langer, 1935; Whipple and Greenstein, 1937) took them up in the years immediately following. Apparently—along with other possible constraints, to be considered in due course—most astronomers and astrophysicists were still maintaining theoretically based conceptions that made further research on extraterrestrial radio emissions seem unpromising. An English cosmologist, D. W. Sciama (1971, p. 50), has since observed that astronomers paid little attention to Jansky because theoretically derived "estimates made at the time led people to believe that our Galaxy would be too weak a radio source to detect." It was not until the end of World War II that English physicists and radar specialists turned their attention to detecting interstellar radio emissions.

Another pattern of preemption of problem choices, this one involving concepts and their associated terms (or labels), was identified in a study of the antecedents of the discovery in 1946 that bacteria reproduce by sexual recombination, a discovery that formed the basis for the specialty of bacterial genetics (Zuckerman, 1974; Lederberg and Zuckerman, 1978). Preemption of this kind occurs through reification, when concepts and the terms designating those concepts fix scientists' imagery of the materials or organisms or processes under investigation, thereby tending to preclude the possibility that selected phenomena of nature could be other than what prevailing concepts and their labels imply they are.

In the case of bacterial recombination, the discovery appears to have been delayed in part by prior concepts that, in effect, shaped biologists' imagery of the mode of reproduction of bacteria. In the nineteenth century, as now, bacteria were taken to be the simplest living organisms. It therefore made sense when the great biological classifier, Ferdinand Cohn, declared in 1872 that bacteria reproduced only by fission and did not recombine sexually (Geison, 1971). According to Cohn, previous reports of bacterial variation (sometimes an indication of sexual mating) were simply erroneous. He went on to *define* bacteria as "chlorophyll-free cells of spherical, oblong, or cylindrical form, sometimes twisted or bent, which multiply exclusively by transverse division" (Geison, 1971, p. 338). To designate this concept, he introduced the term *schizomycetes,* or "fission fungi." For our purpose, the key phrase in that definition is the empirical declaration that bacteria "multiply exclusively by transverse division," thereby ruling out by definition all other forms of reproduction.

This concept of bacteria as simple asexual fungi and its label, *schizomycetes*, became deeply rooted in biologists' thinking. As a result, the modes of bacterial reproduction remained largely unproblematic for about a half century. The overwhelming majority of geneticists could see no reason for looking into the modes of bacterial reproduction, since they "obviously" could not involve any "interesting" genetic processes, which are associated only with sexual reproduction and the crossing of genes. Between 1900 and 1945, only two isolated papers—one by Sherman and Wing in 1937 and the other by Gowan and Lincoln in 1942—specifically explored the question of bacterial sex by genetic methods. Neither of these papers observed the phenomenon of sexual recombination. By accepting the concept of bacteria as asexual organisms, microbiologists and geneticists were diverted from investigating the problem of modes of bacterial reproduction. The concept and its label *schizomycetes*, served temporarily to remove bacterial reproduction from the pool of interesting and researchable problems.

This case also exemplifies the constraining effects on problem choice of accepted scientific law, the same process exemplified by parity conservation and Planck's theory. Cohn's definition of bacteria was linked to Robert Koch's pure culture method, devised in 1881, which, for the first time, provided a usable technique for growing pure—in both the literal and metaphorical senses, *uncontaminated*—cultures of bacteria. When the pure culture method was used, earlier findings, which had seemed to indicate the presence of bacterial variation (alteration in the form of bacteria being investigated), were shown to be false. Apparent "variations" were shown to result from mixtures of diverse strains of bacteria: that is, from contamination in the experiments. Biologists combined the Cohnian concept of bacteria with the Kochian method into what they playfully but respectfully called the "Cohn-Koch Dogma," which stated that "each species of microbe is unchangeable in form and in properties and cannot transform another species" (Henrici, 1934, p. 19). The Cohn-Koch Dogma represented an accumulation of knowledge and was not an idea to be lightly challenged. If Cohn's definition of bacteria and the label *schizomycetes* had not been enough to foreclose the idea of research on bacterial reproduction, the composite Cohn-Koch Dogma, reinforced by its record of otherwise fruitful scientific findings in microbiology, in effect ruled out the problem as illegitimate.

Theoretical concepts, terminology (labels), and scientific laws are only part of the assumption structures that focus the attention of scientists on certain problems and divert it from others. During our studies of selected developments in microbiology, Joshua Lederberg and I noted that experiments in certain fields acquire an unenviable reputation for being *error-prone.* Scientists evidently learn from hard-won experience that the hazards of not easily detected error are far greater in some kinds of experiments than others, even when prevailing standards of procedural caution and accuracy are carefully observed. Studies of bacterial variation and recombination had acquired just such a reputation by the turn of the century. They were, according to one author, "vexatious": messy and very difficult to protect from contamination of various kinds. And because biologists, like other scientists, evidently define research in terms of what Peter Medawar has described as the "art of the soluble," they were understandably motivated to avoid working on bacterial recombination, a problem of dubious respectability and limited prospects.

The rule of thumb that calls for avoiding error-prone fields of inquiry is clearly formulated by Medawar (1967, p. 7) in the opening of his well-known book:

> No scientist is admired for failing in the attempt to solve problems that lie beyond his competence. The most he can hope for is the kindly contempt earned by the Utopian politician. If politics is the art of the possible, research is surely the art of the soluble. . . . Good scientists study the most important problems they think they can solve. It is, after all, their professional business to solve problems, not merely to grapple with them. The spectacle of a scientist locked in combat with the forces of ignorance is not an inspiring one if, in the outcome, the scientist is

> routed. *That is why some of the most important biological problems have not yet appeared on the agenda of practical research* [italics added].

But, of course, and that is the gist of the patterns of preemption that have been examined, prior theoretical assumptions determine what will be defined as soluble and insoluble problems at any given time. A combination of such theoretical and conceptual assumptions apparently preempted research on bacterial recombination for decades. It was not until 1946, in the wake of renewed interest in the biochemical genetics of other simple organisms such as *Neurospora,* that Joshua Lederberg, then a medical student, and Edward L. Tatum, then an established biochemist, first observed sexual recombination in *E. coli.* With this discovery and consequent other ideas and findings, the preemptive Cohnian concept—that bacteria reproduce exclusively by fission—was laid to rest.

Processes of theoretical preemption may help to account for what can be described as "post-mature discoveries"—scientific contributions that presumably could have been made some time before they actually were, if only their narrowly *specific* cognitive ingredients had been sufficient for the outcome.[6] But, as we have seen illustrated by the Jansky discovery of extraterrestrial radio emissions and the Lederberg-Tatum discovery of bacterial recombination, these specifics are embedded in a broader theoretical matrix. That matrix—with its constituent assumptions, concepts, and methods—regulates what will and will not be identified as a salient and soluble problem. Only in retrospect, after the hypothetically "delayed discovery" has actually been made, can it be described as post-mature.

Working scientists often refer to cases of perceived post-maturity. Linus Pauling (1974, p. 771), for example, has stated that "there was no reason why" he should not have discovered the alpha helix eleven years before he and R. B. Corey actually did so. In the domain of quantum electrodynamics, Steven Weinberg (1977, pp. 29–30) states: "All the effects that were calculated in the great days of 1947–49 could have been estimated if not actually calculated at any time after 1934. . . . Why was quantum field theory [the basis for such calculations] not taken more seriously? Observations of this sort suggest that scientists focus on problems they conceive of as soluble but that they do not then and there identify all the soluble problems.[7]

The Jansky case illustrates another and, in this account, concluding form of theoretical preemption in problem choice. Jansky's findings, unlike those of Lederberg and Tatum, did not at once serve as a bridge to further research. Jansky's findings were perceived as an isolated serendipitous discovery rather than an outcome of a concerted research program designed to question some prevailing theoretical and conceptual assumptions. Jansky's work also may have been "neglected" because astronomers were focusing their attention on other problems, which seemed to them more feasible and more consequential. As Edge (1977, p. 331) observes, American astronomers "did not have the requisite skills to adopt radio techniques, and their commitment to a flourishing programme of optical work probably increased their reluctance to change strategy: at a time of economic recession, they were unable to persuade anyone else to take up Jansky's work."[8] Asimov (1975, p. 199) makes a similar observation, attributing the lack of interest in Jansky's findings to the fact that "professional astronomers . . . were caught up in the revolutionary findings of the Hooker reflectors and in the plans for other large optical telescopes and [therefore] shrugged off the microwave finding."

The selective focus of scientists' attention can be thought of in both substantive and formal aspects.[9] Substantively, prior theoretical commitments and technical capabilities direct attention to certain matters as significant and soluble problems; formally, the focusing of scientific attention on certain problems is at the expense of other problems. As Kenneth Burke has observed, "A way of seeing is also a way of not seeing—a focus upon object A involves a neglect of object B" (quoted in Merton, [1940] 1968a, p. 252).

These various patterns of selective scientific attention indicate that the actual research agenda in a given field is characteristically a subset of problems that could, with greater or less difficulty, be worked on. This observation is consistent with the concept of *selective* accumulation of scientific knowledge. That concept refers to short stretches of continuity in scientific development along with hiatus and diversification. (On patterns of "continuities" and "discontinuities" in science, see Merton, 1968c, pp. 8–22.)

If case studies of how scientists come to define certain matters as problematic are at least in their beginnings, the companion question of how they go about selecting problems from the pool of previously identified possibilities has scarcely begun to be comprehensively investigated. The Edge-Mulkay (1976, p. 224 ff.) and the Lederberg-Zuckerman ([1974] 1978) studies found, in the particular cases under review, that two criteria were most frequently used in selecting from arrays of previously identified problems: (1) the assessed scientific importance of a problem (which of course reflects theoretical commitments) and (2) the feasibility of arriving at solutions. Case studies of this sort open up questions requiring study over the spectrum of scientists (though the current feasibility of such comprehensive studies is not self-evident).

Such explicit and tacit cognitive criteria of problem finding, problem relevance, and problem significance could in principle be related to the social as well as the cognitive structure of the sciences under examination. Along with this intricate research agenda is the related question of extrascientific influences upon problem choice. Early investigations along these lines appear in Hessen ([1931] 1971) and Merton ([1938] 1970). (See also Clark, 1937, chap. 6.)[10] In his monograph, Merton rejects both the unqualified externalist position that "the foci of scientific interest" are wholly determined by external (principally economic and military) requirements and the unqualified internalist position that selection of problems for research "is determined by the strict necessity which inheres in each logic-tight compartment of science" (Merton, [1938] 1970, p. 199). In seventeenth-century England, for example, a sizable fraction of general scientific problems was related to extrascientific influences, but these in turn suggested a host of "derivative problems . . . and it is the development of these derivative problems uncovered through continuous scientific study which for the most part accounts for the foci and shifts of attention in given sciences over relatively short periods of time" (Merton, [1938] 1970, p. 49). The interplay of cognitive and social factors in problem choice is all the more complex since judgments about problems worth investigating are also influenced by *social processes internal to science*; for example, by "reactions to the inferred critical attitudes or actual criticism of other scientists and by an adjustment of behavior in accordance with these attitudes" (Merton, [1938] 1970, p. 219).

The generic question of the influence of social processes internal to science upon problem choice has recently been considered in a variety of empirical studies. These include inquiries into the extent to which novice scientists or their mentors determine what the novices are to investigate (Berelson, 1960, p. 178; Crane, 1964, p. 111 ff.) and also into the relative influence of immediate and geographically distant colleagues on problem choice by mature scientists (Glueck and Jauch, 1975). However the process of competition as related to the reward system appears to be attracting the most notice as an influence on problem selection. The social system of science provides institutionalized motivation and reward for achieving priority in solving significant problems at the moving frontier of the field (Merton, [1957] 1973c, chap. 14). This reward system acts to enlarge the numbers of scientists who want to work in the "interesting" and consequential problem areas—the "hot fields," as they are often called—thus intensifying the competition in those areas. Individual scientists then confront a dilemma: Should they work on problems widely defined as interesting and important, and thereby enter into vigorous competition for priority with many others, or should they select other problems involving less competition? (On "population density" and competition in science, see Merton and Lewis, 1971, pp. 157–160; and Hagstrom, 1965, pp. 82–83).

This dilemma is at the core of some of the research by Sullivan, Barboni, and White (1978) on national differences in the foci of attention of particle physicists. They conclude that, in the mode of economic rationality, physicists try to maximize the chances of both achieving priority and solving significant problems. They also emphasize that problem choices are constrained by the experimental technology on hand and access to up-to-date information. Edge and Mulkay (1976, p. 234 ff.) report a similar calculus among radio astronomers, but they go on to observe that the radio astronomers, rather than entering willingly into competition, adopted "a policy of avoiding duplication, and thereby avoiding competition" (p. 237).[11] Such a policy could be adopted during this period of rapid development of radio astronomy, since there were enough important problems to go around. In other papers, these investigators severally hypothesize (Mulkay, 1975, 1976a; Mulkay, Gilbert, and Woolgar, 1975; Edge, 1977) that new problem areas offer scientists a multiplicity of options for research, which tends to reduce competition; at the same time, new important discoveries attract comparatively large numbers of scientists to a limited problem area, which tends to intensify competition.

Without at all implying that social and cognitive processes in science are distinct in anything but an analytical sense, we can note the beginnings of research on cognitive bases of problem choice. The studies by Sullivan and coauthors (Sullivan, White, and Barboni, 1977; Sullivan, Barboni, and White, 1978) are provisional prototypes for one mode of exploring the differential effects of basic contributions on foci of attention. Examining the impact of such contributions as parity nonconservation (1977) and CP (charge conjugation and parity nonconservation) on the physics of weak interactions (1978), they found quite distinct responses by theorists and experimentalists. Theorists moved at once to explore the further implications of these basic contributions, as is indicated in part by their rapidly increasing rate of publication and by concentrations of citations in the theoretical literature. Experimentalists also shifted their attention to problems generated by these contributions, but they did so only after an interval (reflecting familiar real-life constraints upon experimentation, such as obtaining substantial research support and setting up equipment).

Other recent studies attempt to differentiate types of theoretical contributions that have distinct consequences for problem choice. Moravcsik and Murugesan (1977), drawing upon qualitative analysis of citations in papers on theoretical physics, note that some basic theoretical contributions seem to stimulate research, whereas other contributions virtually close off further theoretical work. For the work on nonconservation of parity opened up a whole new class of theoretical ideas, leading physicists to take up a new set of problems and research programs for investigation. In contrast, the Bardeen-Cooper-Schrieffer theory of superconductivity was largely a culminating theory: "After BCS theory, nothing else remained" (Moravcsik and Murugesan, 1977, p. 5).

These several efforts to differentiate types of cognitive contributions and to probe their differential effects on problem choice provide potentials for linking up sociological inquiry with models of scientific development being proposed by philosophers of science.

The foci of attention in the sciences are of course no more than the aggregated problem choices of individuals working in those sciences. Recent studies of changing foci of scientific attention—for instance, studies by Krantz (1965) in the psychology of learning, Goffman (1966) in mast cell research, Goffman and Harmon (1971) in symbolic logic, Sullivan and colleagues (1977, 1978) in the physics of weak interactions, and Gieryn (1977) in astronomy—all show marked short-run variations in attention devoted to particular problem areas, as indicated by changing numbers of publications. These recent findings on bursts of scientific activity in limited problem areas are much the same as those reported by Merton for seventeenth-century British science, based on his classification of papers published in the *Philosophical Transactions* of the Royal Society of London ([1938] 1970, pp. 45–54). He concludes, as already noted, that "short-term fluctuations in scientific interest are primarily determined by the internal history of the science in question" (Merton, [1938] 1970, p. 48). Such an interpretation is altogether consistent with Gerald Holton's (1962) provocative observation that the discovery of new areas of ignorance excites scientists to vigorous activity. In Holton's model, numbers of scientists migrate to these newly identified areas of ignorance; their rate of publication (in these areas) increases rapidly at first and then trails off as the obvious and comparatively more feasible problems are examined. After a time, the immigration of new recruits slows down, and out-migration occurs. (See also Merton and Lewis, 1971, p. 157, on "hot fields"; and Crane, 1972; Mulkay, 1975, 1976a; Mulkay, Gilbert, and Woolgar, 1975.)

The further understanding of problem choice generally and of Holton's hypothesis specifically requires the disaggregation of data on changing foci of attention in specific problem areas into data on shifts in the attention of individual scientists. Garvey and Tomita (1972) in nine disciplines and Gieryn (1977, and in this book) in astronomy have moved in that direction. Each finds changes over the short run in the problems that individual scientists are studying. Such changes, Gieryn's studies tentatively suggest, consist in scientists' adding to and subtracting from their problem sets.[12] Seldom do individuals move entirely from one set of problems to another. Among other things, this means that collective foci of attention in a field can shift rapidly in response to new basic contributions, but individual scientists can still maintain continuity with regard to *some* of the problems they have been working on.

The most striking general impression conveyed by the array of studies dealing with theory choice is that scientists' actual behavior corresponds only imperfectly with epistemological prescriptions of how they should behave. But the gap is still wide between the concepts used by philosophers of science and the crude indicators used in empirical research by sociologists of science. Much more attention to the concept-indicator problem in empirical research will be required before we can determine the ways in which the behavior of scientists accords with and departs from what philosophers of science say ought to obtain in theory choice.

At the same time, studies of problem choice suggest that the actual behavior of scientists corresponds more than one might suppose to often stated methodological precepts which give primacy to assessed significance and feasibility of solution, both deriving from prevailing theoretical contexts. But the studies also find that the selection of general problem areas as well as particular problems is variously influenced by social factors internal to science, such as the extent of competition and expectations of reward, and also by extrascientific factors, such as economic and military needs, which find no place in epistemological or methodological accounts.

Sociologists have just begun to take steps toward describing patterns of theory choice and problem choice for individuals and for aggregates of scientists. These steps are obviously only a prelude to systematic investigation of the cognitive and social sources of decisions that ultimately make for cognitive change in science.

Notes

1. One gathers that Kuhn refers here to choices between paradigms and not to choices between competing theoretical for-

mulations deriving from the same paradigm. On this other kind of theory choice, see Kuhn on "theoretical problems in normal science" ([1962] 1970, pp. 30-34).

2. Interestingly enough, both Kuhn ([1962] 1970, pp. 109–101 and chap. 9) and Merton ([1949] 1968b, pp. 99–100; [1961] 1973d, pp. 65–66) have elected to focus on the phenomenon of "talking past each other" and its various implications. The shared focus probably derives from their both being committed to the basic assumption that the social and cognitive values and the norms of science relate to its cognitive contents (Kuhn, 1977, pp. xxi-xxii; 1977, chap. 13; Merton, 1973e, chaps. 12–13).

3. I also bypass historical investigations attempting to connect theory choice with the philosophical and political orientations of scientists. See, for instance, Forman's (1973, 1974) studies of German physicists' views on indeterminacy in physics and their rejection of mechanistic philosophy.

4. In the everyday idiom of the scientific workplace, a physicist reports that he is measuring "cosmic background radiation," developing "a telescope mirror that compensates automatically for atmospheric distortion," and "search[ing] for quarks with unit charge." A zoologist's catalog of research problems comprises the "lek mating system of the sage grouse," "ecological adaptations of the social structure of blackbirds and wrens," and "the acoustical communication systems of birds." The problems now occupying a cell biologist are described as "the junctions that coordinate the activities of cells in tissues and the structure of photosynthetic membranes." It appears that the scope and specificity of problems that scientists identify in their work vary, but the extent to which they vary has not been examined systematically.

5. Perhaps the most interesting of these failures was Oliver Lodge's experiments, which were apparently hampered by the "electrical noise" surrounding the city of Liverpool even then. In combination with comparatively insensitive instruments, this noise made it impossible for him to detect solar radio waves. He recommended, in fact, that the experiments be duplicated in the country in order to avoid the hazards of urban science. See the short historical account of these failures in Reber (1949), an acknowledged pioneer in radio astronomy, and in Westerhout (1972).

6. The concept of "post-mature discoveries" clearly involves the dangerous but, some of us believe, necessary practice of counterfactual history of a kind implicit in interpretations of histor-

ical sociology. The concept was designed to round out the family of kindred concepts of "premature" and "mature" discoveries as developed in the course of a year's work (1973–74) in the "historical sociology of scientific knowledge" by Yehuda Elkana, Joshua Lederberg, Robert Merton, Arnold Thackray, and Harriet Zuckerman at the Center for Advanced Study in the Behavioral Sciences. Premature contributions are those that, once made, are not immediately followed up and developed by the pertinent community of scientists; scientific significance is attributed to them only later (sometimes after independent rediscoveries). These are, in retrospect, sometimes described as having been "ahead of their time." Mature discoveries are those appearing in their apt time, being recognized and taken up at once; a special subset of such contributions appear in the form of multiple, independent, and more or less *simultaneous* discoveries. On cognitive sources of prematurity, see Stent (1972); on social sources, see Barber (1961); on post-maturity, see Zuckerman (1974); on maturity, see Merton (1973e, chaps. 14–17).

7. In accounting for the fact that physicists did not take quantum field theory seriously, Weinberg (1977) points to theoretical constraints of the general kinds we have identified. He observes that Dirac's theory had "tremendous prestige" and had "worked so well in accounting for the fine structure of the hydrogen spectrum without including self-energy effects" that calculating such effects seemed unnecessary. Moreover, he adds, "the appearance of infinities discredited quantum field theory altogether in many physicists' minds" (p. 30). It had apparently developed an unsavory reputation, like the reputation of ideas about bacterial variation. In related fashion, he adds that physicists have difficulty "bridging the gap" between experiment and theory; theoretical physicists find it hard to realize that their calculations have "something to do with the real world" (p. 30).

8. This may have been true for American astronomers, but there does not seem to have been any intrinsic technical obstacle to pursuing Jansky's findings. In fact, they were followed up by Grote Reber, an amateur astronomer who had had some formal training in astrophysics at the University of Chicago. In the late 1930s and early 1940s, he built equipment that enabled him to begin radio mapping the sky. Reber showed that Planck's blackbody law did not hold for interstellar radiations (Westerhout, 1972, p. 213), and his work finally caught the imagination of astronomers. Otto Struve,

then the editor of the *Astrophysical Journal,* recounts that, after examining Reber's data, he and his colleagues at the Yerkes Observatory began to "realize that a completely new branch of astronomy was emerging" (Struve and Zebergs, 1962, p. 95.)

9. The notions of substantive and formal aspects of preemption in science are compatible with the evolutionary perspective adopted by Toulmin (1967, 1972) and Campbell (1974). Both center on the processes of selection involved in scientists' judgments that some rather than other problems are worth investigating and in their assessment of the significance of scientific contributions. (See Toulmin, 1967, pp. 462–465.)

10. Contemporary work in this genre is represented by Useem's (1976a, 1976b) studies of the extent to which patterns of government research funding affect the choice of problems and procedures. For the sample of social scientists under study, Useem finds that choice of methods more often than choice of problems was affected by patterns of funding.

11. Gaston (1973, p. 167) suggests that there may be cultural differences among scientists. Specifically, British high-energy physicists may be more concerned than their American counterparts to avoid the hurly-burly of competition. (See also Gaston, 1978, pp. 118–121, on the reward systems of the two national science communities.)

12. Several studies report that scientists usually work on several problems at once rather than just one, although the tendency to do so seems to be related to the amount of routine in research (Hargens, 1975, p. 61), to the specific cognitive focus of the laboratory with which scientists are associated (Edge and Mulkay, 1976, p. 298 ff.), and to their being in academic or industrial research (Hagstrom, 1965, p. 161, adapted from Roger Krohn's unpublished dissertation).

References

Agassi, J. *Towards a Rational Philosophical Anthropology.* The Hague: Martinus Nijhoff, 1977.

Asimov, I. *Eyes on the Universe: A History of the Telescope.* Boston: Houghton Mifflin, 1975.

Barber, B. "Resistance by Scientists to Scientific Discovery." *Science,* 1961, *134,* 596–602.

Barber, B. "On the Relations Between Philosophy of Science and Sociology of Science." Discussion paper for Conference on Critical Research Problems, Philosophy of Science Association, Reston, Va., Oct. 1977.

Berelson, B. *Graduate Education in the United States.* New York: McGraw-Hill, 1960.

Bernstein, J. "Profiles; A Question of Parity." *New Yorker,* May 12, 1962, pp. 49 ff.

Campbell, D. T. "Evolutionary Epistemology." In P. A. Schilpp (Ed.), *The Philosophy of Karl Popper.* Vol. 1. La Salle, Ill.: Open Court, 1974.

Carter, G. M. *Peer Review, Citations, and Biomedical Research Policy: NIH Grants to Medical School Faculty.* Santa Monica, Calif.: Rand, 1974.

Clark, G. N. *Science and Social Welfare in the Age of Newton.* Oxford: Clarendon Press, 1937.

Cole, S. "Professional Standing and the Reception of Scientific Discoveries." *American Journal of Sociology,* Sept. 1970, *76,* 286–306.

Cole, S. "The Growth of Scientific Knowledge: Theories of Deviance as a Case Study." In L. Coser (Ed.), *The Idea of Social Structure.* New York: Harcourt Brace Jovanovich, 1975.

Cole, S., Rubin, L., and Cole, J. R. "Peer Review and the Support of Science." *Scientific American,* Oct. 1977, *237,* 34–41.

Crane, D. "The Environment of Discovery." Unpublished doctoral dissertation, Columbia University, New York, 1964.

Crane, D. "The Gatekeepers of Science: Some Factors Affecting the Selection of Articles for Scientific Journals." *American Sociologist,* 1967, *2,* 195–201.

Crane, D. *Invisible Colleges: Diffusion of Knowledge in Scientific Communities.* Chicago: University of Chicago Press, 1972.

Edge, D. O. "The Sociology of Innovation in Modern Astronomy." *Quarterly Journal of the Royal Astronomical Society,* 1977, *18,* 326–339.

Edge, D. O., and Mulkay, M. J. *Astronomy Transformed: The Emergence of Radio Astronomy in Britain.* New York: Wiley-Interscience, 1976.

Forman, P. "Scientific Internationalism and the Weimar Physicists: The Ideology and Its Manipulation in Germany After WWI." *Isis,* 1973, *64,* 151–180.

Forman, P. "The Financial Support and Political Alignment of Physicists in Weimar Germany." *Minerva,* Jan. 1974, *12,* 39–66.

Friis, H. T. "Karl Jansky: His Career at Bell Telephone Laboratories." *Science,* Aug. 25, 1965, *149,* 841–842.

Garvey, W. D., and Tomita, K. "Continuity of Productivity by Scientists in the Years 1968–1971." *Science Studies,* Oct. 1972, *2,* 379–383.

Gaston, J. *Originality and Competition in Science.* Chicago: University of Chicago Press, 1973.

Gaston, J. *The Reward System in British and American Science.* New York: Wiley-Interscience, 1978.

Geison, G. L. "Ferdinand Cohn." In C. C. Gillispie (Ed.), *Dictionary of Scientific Biography.* Vol. 3. New York: Scribner's, 1971.

Gieryn, T. F. "Generation Differences in Research Interests of Scientists." Paper presented at 72nd annual meeting of the American Sociological Association, Chicago, Sept. 1977.

Glueck, W. F., and Jauch, L. R. "Sources of Research Ideas Among Productive Scholars." *Journal of Higher Education,* Jan./Feb. 1975, *46,* 103–114.

Goffman, W. "Mathematical Approach to the Spread of Scientific Ideas: The History of Mast Cell Research." *Nature,* Oct. 29, 1966, *212,* 449–452.

Goffman, W., and Harmon, G. "Mathematical Approach to the Prediction of Scientific Discovery." *Nature,* Jan. 8, 1971, *229,* 103–104.

Grünbaum, A. "Is Falsifiability the Touchstone of Scientific Rationality?: Karl Popper Versus Inductivism." In R. S. Cohen and others (Eds.), *Essays in Memory of Imre Lakatos.* Dordrecht, Netherlands: D. Reidel, 1976a.

Grünbaum, A. "Can a Theory Answer More Questions Than One of Its Rivals?" *British Journal of the Philosophy of Science,* 1976b, *27,* 1–23.

Hagstrom, W. O. *The Scientific Community.* New York: Basic Books, 1965.

Hargens, L. L. *Patterns of Scientific Research: A Comparative Analysis of Research in Three Fields.* Washington, D.C.: American Sociological Association, 1975.

Henrici, A. T. *The Biology of Bacteria.* Lexington, Mass.: Heath, 1934.

Hessen, B. "The Social and Economic Roots of Newton's 'Principia.'" In *Science at the Cross Roads.* London: Frank Cass, [1931] 1971.

Holton, G. "Scientific Research and Scholarship: Notes Toward the Design of Proper Scales." *Daedalus,* 1962, *91,* 362–399.

Howson, C. (Ed.). *Method and Appraisal in the Physical Sciences.* Cambridge: Cambridge University Press, 1976.

Jansky, K. "Directional Studies of Atmospherics at High Frequencies." *Proceedings of the Institute of Radio Engineers,* 1932, *30,* 1920–1932.

Krantz, D. L. "Research Activity in 'Normal' and 'Anomalous' Areas." *Journal of the History of the Behavioral Sciences,* 1965, *1,* 39–42.

Krantz, D. L., and Wiggins, L. "Personal and Impersonal Channels of Recruitment in the Growth of Theory." *Human Development,* 1973, *16,* 133–156.

Kuhn, T. S. *The Structure of Scientific Revolutions.* Chicago: University of Chicago Press, 1962. (Enlarged ed., 1970.)

Kuhn, T. S. "Logic of Discovery or Psychology of Research." In P. A. Schilpp (Ed.), *The Philosophy of Karl Popper.* Vol. 2. La Salle, Ill.: Open Court, 1974.

Kuhn, T. S. *The Essential Tension: Selected Studies in Scientific Tradition and Change.* Chicago: University of Chicago Press, 1977.

Lakatos, I. "Falsification and the Methodology of Research Programmes." In I. Lakatos and A. Musgrave (Eds.), *Criticism and the Growth of Knowledge.* Cambridge, England: Cambridge University Press, 1970.

Lakatos, I. "History of Science and Its Rational Reconstructions." In R. C. Buck and R. S. Cohen (Eds.), *Boston Studies in the Philosophy of Science.* Vol. 8. Dordrecht, Netherlands: D. Reidel, 1971.

Lakatos, I. "Popper on Demarcation and Induction." In P. A. Schilpp (Ed.), *The Philosophy of Karl Popper.* Vol. 1. La Salle, Ill.: Open Court, 1974.

Langer, R. M. "Radio Noises from the Galaxy." *Physical Review,* Jan. 15, 1935, *49,* 209–210.

Latsis, S. (Ed.). *Method and Appraisal in Economics.* Cambridge: Cambridge University Press, 1976.

Lederberg, J., and Zuckerman, H. A. "From Schizomycetes to Bacterial Sexuality: A Case Study of Discontinuity in Science." (Revised draft of a paper presented at the meetings of the American Sociological Association, Toronto, 1974.) Unpublished paper, 1978.

Lindsey, D. *The Scientific Publication System in Social Science: A Study of the Operation of Leading Professional Journals in Psychology, Sociology, and Social Work.* San Francisco: Jossey-Bass, 1978.

Medawar, P. B. *The Art of the Soluble.* London: Methuen, 1967.

Merton, R. K. "Notes on Problem Finding in Sociology." In R. K. Merton, L. Broom, and L. S. Cottrell, Jr. (Eds.), *Sociology Today: Problems and Prospects.* New York: Basic Books, 1959.

Merton, R. K. "Bureaucratic Structure and Personality" [1940]. In R. K. Merton, *Social Theory and Social Structure.* (Enlarged ed.) New York: Free Press, 1968a.

Merton, R. K. "Manifest and Latent Functions" [1949]. In R. K. Merton, *Social Theory and Social Structure.* (Enlarged ed.) New York: Free Press, 1968b.

Merton, R. K. *Social Theory and Social Structure.* (Enlarged ed.) New York: Free Press, 1968c.

Merton, R. K. *Science, Technology and Society in Seventeenth-Century England* [1938]. New York: Harper & Row, 1970. (Reprinted Atlantic Highlands, N.J.: Humanities Press, 1978.)

Merton, R. K. "Multiple Discoveries as Strategic Research Site" [1963]. In R. K. Merton, *The Sociology of Science: Theoretical and Empirical Investigations.* Chicago: University of Chicago Press, 1973a.

Merton, R. K. "The Normative Structure of Science" [1942]. In R. K. Merton, *The Sociology of Science: Theoretical and Empirical Investigations.* Chicago: University of Chicago Press, 1973b.

Merton, R. K. "Priorities in Scientific Discovery" [1957]. In R. K. Merton, *The Sociology of Science: Theoretical and Empirical Investigations.* Chicago: University of Chicago Press, 1973c.

Merton, R. K. "Singletons and Multiples in Science" [1961]. In R. K. Merton, *The Sociology of Science: Theoretical and Empirical Investigations.* Chicago: University of Chicago Press, 1973d.

Merton, R. K. *The Sociology of Science: Theoretical and Empirical Investigations.* Chicago: University of Chicago Press, 1973e.

Merton, R. K., and Lewis, R. "The Competitive Pressures. I: The Race for Priority." *Impact of Science on Society,* 1971, *21,* 151–161.

Moravcsik, M. J., and Murugesan, P. "Citation Patterns in Scientific Revolutions." Unpublished paper, 1977.

Mulkay, M. J. "Three Models of Scientific Development." *Sociological Review,* Aug. 1975, *2,* 509–523, 535–537.

Mulkay, M. J. "The Model of Branching." *Sociological Review,* Feb. 1976a, *24,* 125–133.

Mulkay, M. J. "Norms and Ideology in Science." *Social Science Information,* 1976b, *15* (4–5), 637–656.

Mulkay, M. J., Gilbert, G. N., and Woolgar, S. "Problem Areas and Research Networks in Science." *Sociology,* 1975, *9,* 187–203.

Mullins, N. C. *Theories and Theory Groups in Contemporary American Sociology.* New York: Harper & Row, 1973.

Pauling, L. "The Molecular Basis of Biological Specificity." *Nature,* April 26, 1974, *248,* 769–771.

Polanyi, M. *Personal Knowledge.* London: Routledge and Kegan Paul, 1958.

Popper, K. *The Logic of Scientific Discovery.* New York: Basic Books, 1935. (Enlarged ed., 1959.)

Popper, K. *Conjectures and Refutations.* London: Routledge and Kegan Paul, 1962.

Popper, K. "Autobiography" and "Reply to My Critics." In P. A. Schilpp (Ed.), *The Philosophy of Karl Popper.* Vols. 1 and 2. La Salle, Ill.: Open Court, 1974.

Reber, G. "Radio Astronomy." *Scientific American,* Sept. 1949, *181,* 35–40.

Sciama, D. W. *Modern Cosmology.* Cambridge: Cambridge University Press, 1971.

Stent, G. "Prematurity and Uniqueness in Scientific Discovery." *Scientific American,* Dec. 1972, *227,* 84–93.

Struve, O., and Zebergs, V. *Astronomy of the 20th Century.* New York: Macmillan, 1962.

Sullivan, D., Barboni, E. J., and White, D. H. "Problem Choice and the Sociology of Scientific Competition: An International Case Study in Particle Physics." Unpublished paper, 1978.

Sullivan, D., White, D. H., and Barboni, E. J. "The State of a Science: Indicators in the Specialty of Weak Interactions." *Social Studies of Science,* May 1977, *7,* 167–200.

Toulmin, S. E. "The Evolutionary Development of Natural Science." *American Scientist, 1967, 55,* 456–471.

Toulmin, S. E. *Human Understanding.* Vol. I: *The Collective Use and Evolution of Concepts.* Princeton, N.J.: Princeton University Press, 1972.

Useem, M. "Government Influence on the Social Science Paradigm." *Sociological Quarterly,* Spring 1976a, *17,* 146–161.

Useem, M. "State Production of Social Knowledge: Patterns in Government Financing of Academic Social Research." *American Sociological Review,* Aug. 1976b, *41,* 613–629.

Weinberg, S. "The Search for Unity: Notes for a History of Quantum Field Theory," *Daedalus,* Fall 1977, *106* (4), 17–36.

Weinstein, D. "Determinants of Problem Choice in Scientific Research." *Sociological Symposium,* Summer 1976, *16,* 13–23.

Westerhout, G. "The Early History of Radio Astronomy." *Annals of the New York Academy of Sciences,* Aug. 25, 1972, *198,* 211–218.

Whipple, F. L., and Greenstein, J. L. "On the Origin of Interstellar Radio Disturbances." *Proceedings of the National Academy of Sciences,* 1937, *23,* 177–181.

Whitley, R. D. "The Operation of Science Journals: Two Case Studies in British Social Science." *Sociological Review,* New Series, July 1970a, *18,* 241–258.

Whitley, R. D. "The Formal Communication System of Science: A Study of the Organization of British Social Science Journals." In P. Halmos (Ed.), *The Sociology of Science.* Sociological Review Monograph No. 16. Keele, England: University of Keele, 1970b.

Wynne, B. "C. G. Barkla and the J Phenomenon: A Case Study in the Treatment of Deviance in Physics." *Social Studies of Science,* Sept. 1976, *6,* 307–*348.*

Yang, C. N. "The Law of Parity Conservation and Other Symmetry Laws of Physics." In *Les Prix Nobel en 1957.* Stockholm: Norstedt, 1958.

Zuckerman, H. A. "Cognitive and Social Processes in Scientific Discovery: Recombination in Bacteria as a Prototypal Case." Paper presented at 69th annual meeting of the American Sociological Association, San Francisco, 1974.

Zuckerman, H. A., and Merton, R. K. "Patterns of Evaluation in Science: Institutionalization, Structure and Functions of the Referee System." *Minerva,* Jan. 1971, *9,* 66–100.

5

Thomas F. Gieryn

Problem Retention and Problem Change in Science

A scientist rarely makes a career decision more consequential than the selection of a problem for research. Areas of ignorance in science are variously defined as fundamental or peripheral for the advancement of knowledge. To recognize, to choose for investigation and, in the happy instance, to solve a fundamental problem provides greater satisfaction, derivative prestige and more diverse resources for research than solution of a peripheral problem. Patterns of problem choice have not been accorded much attention in the sociology of science. The handful of systematic analyses are largely found within studies of more encompassing sociological questions—for example, the emergence of scientific specialties.

This chapter moves problem choice in science to center

Note: The research reported in this chapter was supported by a grant from the National Science Foundation (SOC 77-17273) to the Columbia University Program in the Sociology of Science. An early draft was read before the second annual meeting of the Society for the Social Studies of Science, Boston, 1977.

stage. It begins with a brief review of recent sociological thinking on problem choice and contrasts this with a different orientation: the analysis of problem change and problem retention in the careers of scientists. How do scientists differ in the number and variety of their areas of research, the length of time they continue research in an area, and the frequency with which they add new areas? A conceptual formulation for investigating patterns of change and retention in problem choices of individual scientists is reported here. Materials on the problem choice patterns of a sample of American astronomers and astrophysicists illustrate a kind of analysis suggested by this orientation: What social structural conditions make for gradual rather than abrupt changes in the set of problems chosen by a scientist, that is, for the simultaneous occurrence of problem change and problem retention?

A Terminological Proem

Several key concepts in this discussion require definition. The first set of concepts provides descriptions of the units of cognitive structures in science. *Problem area* is defined as the accepted knowledge and recognized questions associated with a substantive object of study or with an instrumentational means of inquiry. A problem area is made up of a number of related though discrete problems, and a number of related problem areas are said to make up a specialty. A "scientific discipline" is defined as a set of related specialties. A hypothetical example from astronomy illustrates this *analytic* hierarchy of units of cognitive structures: "To determine the composition of the atmosphere of Mars" is a scientific problem; it is one of several problems making up the problem area perhaps named "astronomical problems of Mars," which in turn is one of several problem areas making up the specialty "planetary astronomy." The term *specialty* here describes a unit of cognitive structures, although sociologists of science have also used it to describe a unit of the social organization of science.

A second set of concepts provides descriptions of particular behavior patterns of scientists. *Problem choice* is defined as the decision by an individual scientist to carry out a program of research on a related set of problems or, more simply, in a problem area. *Prob-*

lem set is defined as the set of problem areas in which an individual scientist does research at a designated time. *Problem change* and *problem retention* are two patterns in the sequence of problem choices over time; each describes a particular pattern of change and continuity in the problem set.

The unit of analysis for most of this discussion is the problem area, not its constitutive problems, although the term *problem choice* (and its kin: *problem set, problem change,* and *problem retention*) is used instead of the cumbersome but more accurate *problem area choice.* For purposes of empirical investigation, a defining characteristic of a problem choice is taken as the publication of a scientific paper whose subject is within the substantive or technical scope of a problem area.

Three Orientations to Problem Choice in Science

Previous sociological studies of problem choice are largely of three kinds.[1] The first investigates changes in the focuses of research attention among a population of scientists. How do social and cognitive conditions, extrinsic and intrinsic to science, affect the number of scientists working in various disciplines or problem areas? A prototype is found in Robert Merton's analysis of science in seventeenth-century England ([1938] 1978, chaps. 2, 7–10). Although perhaps best known for its discussion of the role of the Puritan ethos in the early institutionalization of scientific activities, that study devotes most of its pages to tracing out changes in the problems chosen for research by English scientists of the times. A quantitative analysis of changing distributions among disciplines and problem areas in that period led Merton to an observation that may hold for twentieth-century science: Changing focuses of research attention derive directly from intrinsic scientific and technical developments and indirectly from scientific concern with extrinsic military, economic, and technological problems. In varying degree, for example, both intrinsic theoretical interest in the velocity of fluids and an extrinsic interest in finding better ways to pump water from mines led Boyle, Hooke, Newton, Halley, and lesser minds to conduct experiments on the compressibility of water.[2]

A second line of inquiry treats problem choice as a collateral issue focusing principally on the emergence of scientific specialties.

What social characteristics and research experiences are common to scientists who choose to enter a certain nascent specialty? Edge and Mulkay (1976) describe the emergence of radio astronomy in Britain in part through descriptions of the intertwined careers of individual scientists who were its earliest or most significant contributors. For example, many radio astronomers were recruited in the latter 1940s and early 1950s from those investigating cosmic rays, a problem area then thought to contain few unsolved, tractable, but significant problems. Other sociological studies focus on noncognitive conditions affecting the decision to begin research in a problem area. For example, Gaston (1973, p. 167) finds that high-energy physicists sometimes avoid problem areas that are highly competitive.[3]

A third orientation to problem choice focuses on identifying sequences of change and continuity in the problem choices of scientists. It is perhaps the rare scientist in any discipline who, having selected a problem area for research, continues to work in that one area throughout his or her career. Most scientists presumably change problem areas, dropping old ones while adding new ones, and many do research in more than one problem area at a time. This raises a question: What social and cognitive conditions make for continuity of research in a problem area (problem retention) and for the start of research in a problem area that is new for the scientist or the cessation of research in a former one (problem change)?

Empirical findings on this question are not easily compared in part because of variability in the scope of cognitive units chosen for analysis, in the indicators of problem choice, and in the definition of change and continuity in sequences of problem choices. The variability is apparent in extracts from several studies which report empirical findings or provide theoretical discussions of what is here called problem retention and problem change:

> We found in 1970–71 that the 2,030 authors [from nine disciplines] in our sample had become divided into two almost equal groups—52 percent were still working in the same subject matter areas as their articles [written two years before], and 48 percent

were no longer working in these areas (Garvey and Tomita, 1972, pp. 379–380).

The question now arises whether individual contributors to the field [of mathematical symbolic logic] tended to move in a pattern among the various major areas of interest. In other words, did individual contributors, upon leaving the subject areas of their immediate interest, tend to move among others according to some specific pattern (Goffman, 1971, p. 183)? [Goffman found that about half of the mathematicians changed "fields" in *successive papers* and that the changes among the seven fields were essentially random.]

Since discipline of training has an inertial effect, the vast majority of researchers stay in a single specialty or work in a related set of specialties throughout their careers. These "stayers" comprise the rank-and-file of active specialties, groups and clusters which are more or less strongly linked in a communication network. . . . "Movers" are scientists who (typically) migrate either short distances—that is, between adjacent specialties—or long distances—that is, between specialties embedded in different disciplines (Chubin, 1976, p. 471). [In summarizing the findings of Krauze (1972), Chubin suggests the possibility that some scientists are more likely to change "specialties" than others.]

Older mathematicians are less likely to change their special fields of interest (Hagstrom, 1965, p. 234). [Hagstrom provides empirical specification for Chubin's hypothesis that the likelihood of changing "fields" is related to professional age. A similar finding is reported by Harmon (1965), who also reports that the likelihood of changing "fields" varies substantially from discipline to discipline.]

When scientists shift from one problem area to another within a single discipline, they tend to shift from areas of low prestige to areas of high prestige. [Then the following footnote:] This would only be true for scientists without established reputations. Distinguished scientists may be able to change special-

> ties and carry their prestige with them (Hagstrom, 1965, p. 53 and note 100). [Allison and Stewart (1974, p. 602) and Mulkay (1972, p. 49) also suggest that the likelihood of changing "problem areas" may be related to the eminence of the scientist.]
>
> Hot fields . . . have a higher rate of immigration and a low rate of emigration, again until they show signs of cooling off (Merton and Lewis, 1971, p. 157). [The authors suggest that the likelihood of "migration" among "fields" is related to the supply of "consequential problems."]

The variety of terms describing units of cognitive structures in science—subject matter areas, subfields, special fields of activity, fields, specialties, subspecialties, and problem areas—suggests that the sociology of science has not yet developed standard categories or even standard measures of the cognitive scope of these various "sectors" of scientific knowledge and practice.

Nevertheless, the empirical findings point to the conclusion that "intellectual migration seems to occur very frequently . . . in most if not all scientific disciplines" (Mulkay, 1972, p. 40). None of the studies found anything particularly problematic about the term *migration* or its several variants: *immigration, emigration, problem switch,* or simply *change.* In everyday parlance, the word *migration* suggests a discrete move from one place to another, and its use often seems to imply that some scientists make frequent migrations from one problem area to another. The migration metaphor suggests that (1) when research in a problem area is begun, earlier work in another area is dropped; (2) there is no overlapping of successive problem choices; (3) scientists work in one problem area at a time. Empirical evidence for the frequency of these patterns is not reported in studies to date, and those making use of the migration metaphor may not intend any of these connotations. To investigate the question further requires a set of concepts allowing for somewhat more fine-grained distinctions among *kinds of changes* in problem sets.

This conceptual formulation begins with the assumption that scientists differ in the *duration of research in a particular problem area* and in the *total number of problem areas in the problem set.* The limiting case of continuity of research in a problem area is one in which a scientist retains an interest throughout his entire career. The pattern for most scientists will no doubt exhibit discontinuities of two kinds: *transient interests,* where a scientist does research in a problem area only during some segment of the career;[4] and *periodic interests,* where a scientist drops research in a problem area for awhile and then returns to it, perhaps repeating this pattern more than once. A scientist who works in more than one problem area could of course have different patterns of continuity or discontinuity for each area. In one of the few empirical studies of duration of interests in problem areas, Mullins (1968) found that the median length of time spent in the "phage group" in biology was about six years and that few scientists continued to work on these problems for as long as ten years.

Isaiah Berlin (1953) has revived an observation of the Greek poet Archilocus: "The fox knows many things, but the hedgehog knows one big thing." (For an application of the aphorism see Jahoda, 1978.) That observation is consonant with several recent studies that find variation in the size of problem sets of scientists. It should be emphasized once again, however, that differences in the defined scope of problem areas make it difficult to compare the reported findings. Edge and Mulkay (1976, p. 298) found that the social structure of research centers affects the size of problem sets of astronomers. At the Cambridge radio astronomy unit, only four of twelve scientists had 70 percent or more of their articles concentrated in one "subject area," while at the Jodrell Bank site (Manchester) over half had such concentrated patterns of problem choice. Hargens (1975, p. 61) found differences among scientific disciplines: "Chemists tend to carry on research in several questions simultaneously to a greater extent than do mathematicians and political scientists."

Variation in the duration of research in a single problem area and in the number of problem areas in which a scientist does research suggests the possibility of various kinds of changes in the sequence of problem choices. One scientist might add a new interest in a problem area while retaining all other interests; another might substitute one problem area for an earlier one while at the same time continuing research in several other areas; a third might exchange a former problem set for a completely new one. Table 1 sets out a typology of the six logically possible kinds of changes in the problem set. The letters "a" through "f" indicate research in a distinct problem area; the number of problem areas at each observation is assigned arbitrarily.

Duplication is the limiting case of problem retention: research in all problem areas is continued from the first time of observation to the second. Migration is the limiting case of problem change: earlier interests are replaced by an entirely new set. Accretion, selective disengagement, and selective substitution represent differing patterns of simultaneous problem retention and problem change: research in some problem areas is continued while work in others is either added or dropped. Since the typology applies only to comparisons between problem sets at two times of observation, it cannot of course serve to summarize sequences of problem choice throughout a professional life. For example, a series of accretions may be followed later in the career by successive years of disengagement and then withdrawal.

An obvious question is suggested by this typology: What is the relative frequency of each kind of change? Is migration, defined here as a complete turnover in the problem set, more or less frequent than gradual changes, which involve simultaneous problem change and problem retention? Or, more specifically, will the incidence of designated patterns of problem choice vary for different disciplines at different phases of development, or for scientists of different professional age or eminence? In such analyses, empirical findings will in part reflect the defined scope of problem areas and the length of time between observations. Expanding the scope

Table 1. Changes in the Problem Set

Type of Change	*Problem Set At Time 1*	*Problem Set At Time 2*
Duplication	abc	abc
Accretion	abc	abcd
Selective substitution	abc	abd
Migration	abc	def
Selective disengagement	abc	ab
Withdrawal from research	abc	none

of problem areas to include increasingly variegated problems will of course tend to decrease the observed frequency of problem change; expanding the time between successive observations will tend to increase measured problem change. Still, beyond these not easily overcome measurement difficulties, there is theoretical reason to suppose that gradual change in problem sets (accretion, selective disengagement, and selective substitution) is more common than complete problem retention (duplication) *or* complete problem change (migration). The organization of the reward and opportunity structure of science, it will be suggested, encourages scientists to continue research in some problem areas even as they add or delete research in others.

Patterns of problem choice are of course influenced by developments in scientific knowledge and research methods. Developments in the cognitive structure of science variously make for problem retention and problem change. Yellin (1972) has provided a useful conceptual framework for investigating the impact of "excitements"—discoveries, new theories, and technical breakthroughs—on focuses of research attention in a discipline. While recognizing the importance of such cognitive conditions and the need for investigating them further, I shall focus on identified aspects of the social structure of science that make for problem retention and problem change. Certain elements of the reward and opportunity structure, and of the normative structure of science, temper the impact of cognitive and technical developments on patterns of problem choice.

Change and Continuity in the Problem Set

Addition of a new interest in a problem area, it is hypothesized, will generally be accompanied by retention of research in one or more other problem areas. Termination of research in a problem area (except in the case of full withdrawal from research) will probably be preceded by a substantial period in which the subsequently terminated research and the perpetuated research exist together. Thus, complete substitution of one problem set for a completely different one is expected to be a less common occurrence than simultaneous problem change and problem retention.

There are several reasons to suppose that this is the case for most aggregates of scientists.

First, initiating research in a problem area is a riskier venture than continuing work in an old one. The chances of failure of research in a new problem area are presumably greater. *Novices* (a term not necessarily associated with professional age) will tend to have less familiarity with required technical skills and less awareness of the new critically significant problems. Nor have they as yet made the associations that would allow them to get helpful advice from other specialists. If scientists continue with earlier research in a problem area, they will have an ongoing line of inquiry to fall back on if a new venture fails. (The same observation is made by Hagstrom, 1965, p. 161, and Crane, 1965, p. 707.)

Second, palpable results from new research in a problem area will not be immediate.[5] Some scientists at certain times in their career may need to maintain an uninterrupted production of scientific papers, even while they begin research in a problem area new to them; continuation of research in other problem areas allows for this. Also, accumulation of research experience in a problem area may mean that less preparatory time is required to produce each additional paper on these problems. The first paper worked out in a problem area will probably require more preparation time than the tenth. Retention of long-standing interests in a problem area may make it easier for a scientist to satisfy motivations and social expectations to make contributions to the scientific literature. These concerns are perhaps reflected in the following exchange between the astrophysicist S. Chandrasekhar and Spencer R. Weart during an oral history interview (May 19, 1977, p. 99; from copy of transcript at the Niels Bohr Library, American Institute of Physics, New York):

> *C*: In fact you raise an important question. One of the fears I always had for many years was whether I could continue generating scientific problems, nontrivial problems, for long periods of time. I sort of felt that if one gave up doing serious scientific research for a period, then one might not be able to get back to it. And so, in order to essentially protect myself

against losing a grip on science, or somehow stopping the flow of ideas, I kept on.

W: I see. Is this also why your work tends to overlap—even after you finish a book, there will be a certain momentum and you'll continue producing a few papers? While you're starting the next one?

C: That's right. I make an overlap. I gradually terminate one, while picking up something else.

W: Till the momentum gets going?

C: Yes.

Third, even when new research in a problem area begins to have palpable results, a scientist may continue with earlier research that makes use of accumulated skills and resources. Only some part of the trained capacity for research in earlier problem areas may be transferable for research in the new one. This tendency is implied in the Weinberg (1967, p. 40) question: "To what extent has science suffered because scientists are understandably reluctant to junk a very expensive piece of apparatus, even though the logical force of scientific development would send them in other directions that do not use this particular apparatus?"

Fourth, institutionalized patterns for evaluating the role performance of scientists create certain advantages for scientists who do research for a substantial time in a particular problem area. If scientists are recognized members of the network of researchers who work in a problem area, they can call upon that network to provide informed assessments of past contributions when such assessments are required, say, by funding agencies or departments considering the granting of tenure. This advantage may be less often available to scientists who do not maintain a long-standing program of research in a problem area and who may have a more difficult time becoming part of an established research network.[6]

Fifth, even when scientists would prefer to direct their attention to new sets of problems, the structurally patterned distribution of research opportunities may encourage retention of the old problem areas. After doing much work in a problem area, scientists may come to be recognized by peers as having a special interest and competence in these problems. These social perceptions of their research by peers will not necessarily correspond to their current interests. The collective imputations of problem choices take on special importance when they are shared by scientists who occupy strategic statuses in the system for allocating resources for research: the department chairman with a vacant teaching or research position searches for candidates perceived as doing research in a certain problem area; peer review committees that allocate research grants or fellowships consider lines of research experience and accomplishments in comparing proposals for research in problem areas (as Cole, Rubin, and Cole, 1977, p. 40, found in their study of the National Science Foundation); the editor of a topical volume (or the organizer of a session at a professional meeting) solicits contributions from scientists with demonstrated accomplishments in a problem area; and the departmental colleague looks for collaborators who bring needed trained capacities and research interests. In some cases, the problem choices imputed by others are identical with current self-defined problem choices, and the provision of opportunities to pursue a set of problems is matched by the motivation to do so. In other cases, "opportunities" may be provided for research that once interested a scientist but now have no place on the scientist's agenda. This can lead to a self-fulfilling prophecy: Incorrect assumptions about another's current problem choices lead to providing opportunities for types of research that were once of interest but are no longer. Some scientists will not find it in themselves to turn down these opportunities, and so renew former interests in a problem area even as they try to initiate new ones.

Sixth, the hypothesized prevalence of simultaneous problem retention and problem change is also taken to reflect a coexisting norm and counternorm that in effect call for certain patterns of problem choice. Merton's ([1963] 1976, pp. 33–34) list of opposing norms which constrain the behavior of scientists can be augmented:

> A scientist should not be excessively parochial by choosing to work only on a very narrow range of problems. *But* he should not become a dilettante by skittering from one problem area to another without adequate preparation. (Compare Max Weber's remark that only "by strict specialization can the scien-

> tific worker [achieve] something that will endure" ([1918] 1946, p. 135) with what Fermi said of his own career: "change fields every five years" (in Weinberg, 1967, p. 135) or with what Weber ([1918] 1946, p. 135) himself observed almost in the same breath as the above: "many of our very best hypotheses and insights are due precisely to dilettantes."

Opposing expectations imbue the selection of research problems with a kind of sociological ambivalence, best satisfied by compromise: scientists should from time to time expand their problem set; at the same time, they should continue research in former problem areas for at least the period needed to master skills or instruments required for research in the new problem area.

If these conjectures have any merit, one should find a prevalence of gradual rather than abrupt changes in scientists' problem sets. Problem change (addition of new research in a problem area or cessation of earlier research) should generally be accompanied by problem retention (the continuity of research in one or more other problem areas).

Problem Choice in Astronomy: An Illustration

The empirical examination of this hypothesis faces a discouraging array of measurement difficulties. Some of these will emerge in this brief report, which examines selected patterns of problem choice among a sample of 2,308 American astronomers and astrophysicists. The list of astronomers was generated from disciplinary listings in the thirteenth edition of *American Men and Women of Science* and from recent membership directories of the American Astronomical Society. The latter source in particular identifies scientists in collateral disciplines—physics, mainly, but also geology, mathematics, and chemistry—who work on astronomical problems. Only scientists who had published at least one scientific paper between 1951 and 1975 were included.

Astronomy and Astrophysics Abstracts was used to identify the problem choices of individual scientists.[7] This source biennially sorts the world astronomical literature into ninety-three subject categories. Each scientist was identified in the author index of volumes covering 1951 through 1975, and the number of his or her publications in each category for each year was recorded.

To determine empirical frequencies of the six patterns of change in the problem set, the publications of a sample of scientists were examined first in the 1963–1965 triennia and again in the 1973–1975 triennia. The subject categories in the *Abstracts* were collapsed into eleven broader units, taken to represent major problem areas in astronomy. As an example, in the 1973–1975 period Chandrasekhar published fifteen papers; twelve are classified by the *Abstracts* in categories comprising the problem area of theoretical astrophysics, in stellar astronomy, and one in celestial mechanics. Chandrasekhar is said to be doing research in these three problem areas in 1973–1975.[8]

Table 2 reports the empirical distribution of scientists among the six types of changes in the problem set. In this illustrative analysis, only scientists who were doing research in *two* problem areas in 1963–1965 were included. Thirteen of the 125 scientists did not publish a scientific paper in 1973–1975. Of the remaining 112 scientists, 99 changed problem choices between the two periods. Accretion is the most common type of change, while migration—a complete turnover of problem choices—is the least common. Of the 79 scientists who added one or more new problem areas (the sum of accretion, selective substitution, and migration), 66 (83.5 percent) retained at least one of their former areas of research, while 13 (16.5 percent) apparently dropped earlier research while beginning something new.

The sketchy evidence is consistent with the hypothesis that gradual changes in the problem set are more common than abrupt

Table 2. Changes in the Problem Sets of Selected Astronomers

Type of Change	*Number and Percentage of Scientists*	
Duplication	13	10.4%
Accretion	41	32.8
Selective substitution	25	20.0
Migration	13	10.4
Selective disengagement	20	16.0
Withdrawal from research	13	10.4
	125	100.0%

ones, but it is hardly conclusive. As I have noted, the results are much affected by the cognitive scope of astronomical problem areas. What will be the findings as the scope of problem areas is changed, increasing or decreasing the range and number of problems? How would the distribution change if, say, the fine-grained categories provided by the *Abstracts* are defined as problem areas? Also, the findings will change if the time intervals between observations or the duration of the period of observation (now three years) is varied. Perhaps different results will emerge if scientists working in only one problem area or in more than two in 1963–1965 are included in the sample under study. Finally, other sources of information on problem choices could be adopted as complements for this use of scientific publications and their classification by the *Abstracts*. Do scientists' own perceptions of their problem choices (as elicited, for example, by questionnaires) differ from patterns indicated by their publications?

Moreover, other kinds of analyses are suggested by the same hypothesis. For example, if problem retention generally accompanies new problem choices, then one should find that older scientists work in more problem areas than younger scientists. One should also find (as Stehr and Larson, 1972, and Gieryn, 1977, have found) age-related differences in foci of research attention. Older scientists may retain interests in problem areas less often chosen for research by younger cohorts of scientists.

Conclusion

This study of social structural conditions making for problem retention and problem change is only another step in the still barely explored subject of problem choice in science. It does not consider the sources and consequences of differing durations of research in a problem area (for individual scientists or for science as an institution); it does not consider the consequences of an expansion of the number of problem areas pursued at the same time by scientists; it does not examine how these various patterns of change and continuity differ among scientific disciplines with their possibily different social and cognitive structures; and it only hints at the biographical characteristics (professional age, eminence) that may be related to incidence of one or another of these patterns of problem choices.

The typology of changes and continuities in problem choices may offer a useful context to explore these and related questions.

Notes

1. Chapter Four by Harriet Zuckerman in this collection provides a detailed summary of investigations of problem choice and theory choice in science.

2. Problem choice and focuses of research attention have been enduring themes in Merton's work in the sociology of science. See especially Merton (1959) on "problem finding" and Zuckerman and Merton (1972) on age-stratified patterns of problem choice. For other treatments of the subject, see Stehr and Larson (1972) on generation differences in focuses of attention among sociologists; and Price (1963), Menard (1971), and Crane (1972) on variations in rates of growth in various scientific disciplines and problem areas.

3. The substantial number of studies of scientific specialties—which in varying fashion consider problem choice as an analytical issue—is reviewed by Chubin (1976) and by Edge and Mulkay (1976, pp. 359–394). Other discussions of the relationship between competition and problem choice are provided by Hagstrom (1965, chap. 2; 1974) and Merton ([1957] 1973, chap. 14).

4. Price and Gürsey (1976) define "transient authors" as those whose names appear in the scientific literature for only one year.

5. The amount of research time required to begin producing palpable results will probably vary greatly among disciplines and problem areas. A psychologist reported to Crane (1965, p. 706): "There are no substantial contributions of the hit-and-run type. Substantial contributions come from ten to twenty years of working on basically one problem area." A chemist told Hagstrom (1965, p. 160): "It takes at least one or two, sometimes up to five, years to get to the publication stage." Presumably, scientists in disciplines with more codified cognitive structures would be able to make contributions after a shorter period of research in a problem area; one might also expect that problem change will be greater in disciplines with well-codified cognitive structures. On codification, see Zuckerman and Merton (1972, p. 302) and initial empirical findings in Cole, Cole, and Dietrich (1978, pp. 209–251).

6. The visibility of scientists has been investigated by Cole and Cole (1973). They conclude that "the work of physicists in each of these specialties is more visible to their colleagues in the same specialty than to colleagues outside it . . . [but] the differences are

considerably smaller than expected" (p. 172). They also report that "physicists in any one specialty have relatively high awareness of high quality work in all specialties" (p. 172). This same degree of visibility may not extend to the work of bench scientists, as Ziman (1968, pp. 61–62) seems to suggest: "There are, of course, always a few powerful eclectic minds who will attempt the synthesis, but until they have achieved sufficient personal status they do not find a ready audience for their work. It is much easier to join a specialty . . . than to create interests embracing a number of these little villages of the mind."

7. The difficulties in the use of scientific abstracts for sociological purposes are legion. Woolgar (1976) provides an informative critique of their use, apposite for this chapter because it takes astronomical research on pulsars as an example. An examination of the *Astronomy and Astrophysics Abstracts* is provided by Gieryn (1978, chap. 2). Several studies have made effective use of abstracts for determining problem choices: Menard (1971); Crane (1972); and Sullivan, White, and Barboni (1977).

8. The Chandrasekhar example incidentally raises an important issue: Some problem areas clearly receive greater attention than others in the problem set of individual scientists. The distribution of publications of a scientist among problem areas may permit investigators to distinguish between problem areas of central concern and those of peripheral concern.

References

Allison, P. D., and Stewart, J. A. "Productivity Differences Among Scientists: Evidence for Accumulative Advantage." *American Sociological Review,* 1974, *39,* 596–606.

Berlin, I. *The Hedgehog and the Fox.* New York: Simon & Schuster, 1953.

Chubin, D. "The Conceptualization of Scientific Specialties." *Sociological Quarterly,* 1976, *17,* 448–476.

Cole, J. R., and Cole, S. *Social Stratification in Science.* Chicago: University of Chicago Press, 1973.

Cole, S., Cole, J. R., and Dietrich, L. "Measuring the Cognitive State of Scientific Disciplines." In Y. Elkana and others (Eds.), *Toward a Metric of Science: The Advent of Science Indicators.* New York: Wiley, 1978.

Cole, S., Rubin, L., and Cole, J. R. "Peer Review and the Support of Science." *Scientific American,* Oct. 1977, *237,* 34–41.

Crane, D. "Scientists at Major and Minor Universities: A Study of Productivity and Recognition." *American Sociological Review,* 1965, *30,* 699–714.

Crane, D. *Invisible Colleges: Diffusion of Knowledge in Scientific Communities.* Chicago: University of Chicago Press, 1972.

Edge, D. O., and Mulkay, M. J. *Astronomy Transformed: The Emergence of Radio Astronomy in Britain.* New York: Wiley-Interscience, 1976.

Garvey, W. D., and Tomita, K. "Continuity of Productivity by Scientists in the Years 1968-1971." *Science Studies,* Oct. 1972, *2,* 379–383.

Gaston, J. *Originality and Competition in Science.* Chicago: University of Chicago Press, 1973.

Gieryn, T. F. "Generation Differences in the Research Interests of Scientists." Paper presented at 72nd annual meeting of the American Sociological Association, Chicago, Sept. 1977.

Gieryn, T. F. "Problem Choice in Science: American Astronomers, 1951–75." Unpublished doctoral dissertation, Columbia University, New York, 1978.

Goffman, W. "A Mathematical Model for Analyzing the Growth of a Scientific Discipline." *Journal of the Association for Computing Machinery,* 1971, *18,* 173–185.

Hagstrom, W. O. *The Scientific Community.* New York: Basic Books, 1965.

Hagstrom, W. O. "Competition in Science." *American Sociological Review,* 1974, *39,* 1–18.

Hargens, L. L. *Patterns of Scientific Research: A Comparative Analysis of Research in Three Fields.* Washington, D.C.: American Sociological Association, 1975.

Harmon, L. R. *Profiles of Ph.D.'s in the Sciences.* Washington, D.C.: National Academy of Sciences, 1965.

Jahoda, M. "Paul F. Lazarsfeld: Hedgehog or Fox?" In R. K. Merton, J. S. Coleman, and P. H. Rossi (Eds.), *Qualitative and Quantitative Social Research: Papers in Honor of Paul F. Lazarsfeld.* New York: Free Press, 1978.

Krauze, T. K. "Social and Intellectual Structures of Science—A Mathematical Analysis." *Science Studies,* 1972, *2,* 369–378.

Menard, H. *Science: Growth and Change.* Cambridge, Mass.: Harvard University Press, 1971.

Merton, R. K. *Science, Technology and Society in Seventeenth-Century England* [1938]. Atlantic Highlands, N.J.: Humanities Press, 1978.

Merton, R. K. "Notes on Problem Finding in Sociology." In R. K. Merton, L. Broom, and L. S. Cottrell, Jr. (Eds.), *Sociology Today: Problems and Prospects.* New York: Basic Books, 1959.

Merton, R. K. "Priorities in Scientific Discovery" [1957]. In N. W. Storer (Ed.), *The Sociology of Science.* Chicago: University of Chicago Press, 1973.

Merton, R. K. "The Ambivalence of Scientists" [1963]. In R. K. Merton, (Ed.), *Sociological Ambivalence and Other Essays.* New York: Free Press, 1976.

Merton, R. K., and Lewis, R. "The Competitive Pressures. I: The Race for Priority. *Impact of Science on Society.* 1971, *21,* 151–161.

Mulkay, M. J. *The Social Process of Innovation: A Study in the Sociology of Science.* London: Macmillan, 1972.

Mullins, N. C. "The Distribution of Social and Cultural Properties in Informal Communication Networks Among Biological Scientists." *American Sociological Review, 1968, 33,* 786–797.

Price, D. J. de S. *Little Science, Big Science.* New York: Columbia University Press, 1963.

Price, D. J. de S., and Gürsey, S. "Studies in Scientometrics. Part I: Transience and Continuance in Scientific Authorship." *International Forum on Information and Documentation,* 1976, *1,* 17–24.

Stehr, N., and Larson, L. E. "The Rise and Decline of Areas of Specialization." *American Sociologist,* 1972, 7 (3), 5–6.

Sullivan, D., White, D. H., and Barboni, E. J. "The State of a Science: Indicators in the Specialty of Weak Interactions." *Social Studies of Science,* May 1977, *7,* 167–200.

Weber, M. "Science as a Vocation" [1918]. In H. Gerth and C. W. Mills (Eds.), *From Max Weber.* New York: Oxford University Press, 1946.

Weinberg, A. *Reflections on Big Science.* Cambridge, Mass: M.I.T. Press, 1967.

Woolgar, S. "The Identification and Definition of Scientific Collectivities." In G. LeMaine and others (Eds.), *Perspectives on the*

Emergence of Scientific Disciplines. The Hague, Netherlands: Mouton, 1976.

Yellin, J., "A Model for Research Problem Allocation Among Members of a Scientific Community." *Journal of Mathematical Sociology,* 1972, *2,* 1–36.

Ziman, J. *Public Knowledge.* Cambridge, England: Cambridge University Press, 1968.

Zuckerman, H., and Merton, R. K. "Age, Aging, and Age Structure in Science." In M. W., Riley, M. Johnson, and A. Foner (Eds.), *Aging and Society.* Vol. 3: *A Sociology of Age Stratification.* New York: Russell Sage Foundation, 1972.

Part 3

Different Approaches

The chapters in this section provide an overview of the state of the art in contemporary sociology of science, suggest ways that sociology and history might profit from limited interaction on questions of mutual interest, show that macro-level analyses of science have many implications for sociological explanation, and suggest some reasons why the different approaches in the sociology of science themselves are not independent of sociological influences.

Lowell Hargens demonstrates how method and theory have interacted to advance the state of knowledge; he shows clearly their reciprocal relationship, which more often is recognized in the abstract than in the specific. To some sociologists the concept of progress represents an ideology that is either dangerous or intellectually dishonest (invalid). Be that as it may, Hargens, without implying progress, shows that the sociology of science has benefited from new developments in method and theory. The result is well known: The state of knowledge in the sociology of science is much greater today than it was twenty years ago. Could that be progress?

Donald Beaver examines the positive aspects of potential interaction between historians and sociologists of science. He does not suggest, however, that the two perspectives should be fused. Indeed, his frank and logical discussion of the resistance to interaction should be required reading for all the university and governmental bureaucrats who, seeing that disciplinary perspectives are by definition limited, believe that forced marriage into some interdisciplinary structures will miraculously be able to take a dependent variable and explain 100 percent of its variation. Beaver's discussion of the intellectual and the social bases for desirable interaction—but interaction only on a limited scale—may be seen by some as overcautious. Others may consider it an accurate explanation of reality.

Derek Price's short chapter is pregnant with ideas, implications, and trends to be explained. Historians may dislike the data; sociologists may dislike the historical perspective. In either case, this chapter is not likely to evoke a mild response. Instead, ideally and characteristically, it should give rise to considerable debate over the use of data aggregated in this way and over the interpretation of the data. Even if there is a consensus on the relevance and validity of the data, the implications promise to be interesting.

As the sociology of science became accepted more, and as more scholars began to pursue research on the problems, other perspectives began to arise and be discussed. Thomas Kuhn's (1962) book *The Structure of Scientific Revolutions* provided what some came to see as an alternative framework for the sociology of science. Moreover, Kuhn's ideas were rather simple to understand, and his examples were vivid, so that, in addition to the consensus that his work is an important contribution to the history and sociology of science, Kuhn's ideas diffused very rapidly.

For some scholars, Kuhnian historical sociology of science became a kind of obvious anti-Mertonian sociology of science. Apparently, it is of no consequence that Merton (see Merton, 1977) and Kuhn (see Kuhn, 1977) do not agree with such claims about differences in their ideas. Thus, the debate has continued without the participation of the main characters.

There are, of course, many other developments in the sociology of science that are not directly connected, in any explicit manner, to the problem of reconciling the models of science developed by Merton and Kuhn (for literature reviews and bibliographies, see Barber, 1959; Kaplan, 1964; Ben-David and Sullivan, 1975; for a review of the literature on development of specialties in science, see Chubin, 1976). In any case, it is not unreasonable to suggest that Merton's and Kuhn's ideas are complementary. The former describes the way that science operates generally but does not include a description of the way that science changes its content; the latter describes the largely social process whereby one set of perspectives is exchanged for a new set.

Merton's influence in the sociology of science is unquestioned, and one of his most important contributions is his formulation of the ethos of science, comprised of norms that specify the social relations in science. This formulation provided the basis for research on several fronts, but especially on the nature of the reward and communication systems in science. Research on the reward system is one of the most feasible types of research in sociology generally, and in the sociology of science particularly. The criticism of Mertonian sociology of science generally focuses on this aspect.

Nico Stehr summarizes some of this criticism, dividing the issues into four categories of questions. He argues that the discussion about the norms of science fails to reach consensus because of the presuppositions and the perspective taken by those participating in the debate. As a participant in this debate, I believe that Stehr has selected important issues from the discussion and that he has removed the debate to a level of discourse higher than previously has been available (for my views about these criticisms, see Stehr's references).

Joseph Ben-David attempts to explain a clearly observable phenomenon: the general preference for Mertonian sociology in the United States and the more nearly unanimous preference in Great Britain for other-than-Mertonian sociology of science. In contrast to Stehr's cognitive approach (in Chapter Nine), Ben-David's approach is a more purely sociological one. He argues that the American and British preferences are not chauvanistic choices but the consequences of the social context in which the researchers are located and in which the research is done. Ben-David's chapter thus serves as an appropriate conclusion to this volume. It summarizes accomplishments in the field and suggests why the theory and problem choices of scholars have brought us to this point.

References

Barber, B. "The Sociology of Science." In R. K. Merton and others (Eds.), *Sociology Today.* New York: Basic Books, 1959.

Ben-David, J., and Sullivan, T. A. "Sociology of Science." In A. Inkeles and others (Eds.), *Annual Review of Sociology.* Palo Alto: Annual Reviews, 1975.

Chubin, D. "The Conceptualization of Scientific Specialties." *The Sociological Quarterly,* 1976, *17,* 448–476.

Kaplan, N. "The Sociology of Science." In R. E. Faris (Ed.), *Handbook of Modern Sociology.* Chicago: Rand McNally, 1964.

Kuhn, T. *The Structure of Scientific Revolutions.* Chicago: University of Chicago Press, 1962.

Kuhn, T. *The Essential Tension: Selected Studies in Scientific Tradition and Change.* Chicago: University of Chicago Press, 1977.

Merton, R. K., "The Sociology of Science: An Episodic Memoir." In R. K. Merton and J. Gaston (Eds.), *The Sociology of Science in Europe.* Carbondale: Southern Illinois University Press, 1977.

Merton, R. K., and Gaston, J. (Eds.). *The Sociology of Science in Europe.* Carbondale: Southern Illinois University Press, 1977.

6

Lowell Hargens

Theory and Method in the Sociology of Science

The sociology of science has recently been treated as an exemplification of general structural and developmental characteristics of all scientific specialties (Cole and Zuckerman, 1975; Merton, 1977), but it may also be used as a microcosmic model of sociology (for discussions of the varying levels of correspondence between disciplines and their constituent specialties, see Hagstrom, 1964, pp. 194–195; Hargens, 1975, pp. 5–16; Whitley, 1976). With respect to research techniques, both specialty and discipline are characterized by substantial gaps between theory and method. That is, the procedures of data collection, measurement, and analysis are not so closely related to substantive theory in these areas as they are in many of the natural sciences (Kuhn, 1961; Knorr, 1975). The continuous appearance of essays criticizing the foundations of commonly used research techniques is symptomatic of the disjunction between theory and method in sociology and it also characterizes controversy in the sociology of science (for examples see respectively Cicourel, 1964, and Edge, 1977).

Two general methodological conditions are associated with

Note: This chapter benefited from the comments and suggestions of Nicholas C. Mullins and Barbara F. Reskin.

the kind of disjunction between theory and method found in sociology and the sociology of science. The first is that concrete research techniques are employed generally, rather than being more or less specific to individual research areas; indeed, research techniques favored by sociologists are widely employed by members of other social science disciplines. When research techniques are not specific to research areas, training in their use is likely to be provided in general courses on methodology; in contrast, in fields with greater specificity, students are more likely to learn research techniques during the actual performance of research (Whitley, 1977). In addition, major innovations in research techniques are less likely to come from practitioners in substantive research areas and more likely to consist of methods being borrowed from other disciplines. In these instances the researchers responsible for introducing the innovations are more accurately described as transmitters and popularizers than as innovators.

A second methodological condition prevalent in fields with weak links between theory and method is that researchers tend to maintain a pluralistic stance about methodological issues that is consistent with their theoretical pluralism on substantive issues (Hagstrom, 1964; Cole, 1975, p. 208). This pluralism is permeated by a skepticism (usually instilled during graduate training) about the adequacy of existing research techniques and a similar skepticism about the adequacy of existing theoretical frameworks (Kuhn, 1970, p. 164). The pluralistic stance is supported by attempts to outline decision rules for the selection of research techniques appropriate to different types of substantive problems (see, for example, Zelditch, 1962) and by advocacy of "triangulation" (that is, the use of several techniques simultaneously, in order to offset the inevitable weakness and bias of any one method (for example, see Webb and others, 1966; Mulkay, 1974). Such methodological advice tempers the skepticism, which, if unchecked, threatens the autonomy of a field (Hagstrom, 1965, pp. 254–286; Lammers, 1974).

That there are substantial gaps between theory and method in sociology does not imply that theory and research are independent of one another; it means only that their relationship is usually a very loose one. Theoretical formulations often are not amenable to empirical investigation because appropriate research techniques are unavailable; therefore, these formulations receive no empirical attention until after new techniques are adopted. In many instances, the availability of techniques influences the lines of inquiry followed by researchers more than current theoretical work does. But research techniques have extremely general applicability and are often described as being "contentless" because they lack close ties with substantive concerns. As a result of these conditions, ongoing research typically bears only a thematic relationship to existing theory (Cole, 1975). These characteristics of research in sociology and the sociology of science contrast sharply with the nature of research in fields that show a more intimate relationship between theory and method (Kuhn, 1970, pp. 24–32).

In the following pages, I illustrate these points about research techniques in fields with weak links between theory and method by briefly examining two sets of techniques—causal models and methods of network analysis—that have been used extensively in the sociology of science during recent years. I emphasize the ways these techniques have interacted with substantive concerns rather than their formal methodological characteristics. After considering these two sets of techniques, I return to the more general question of relations between theory and method in the sociology of science.

Causal Models and Social Stratification in Science

During the early and middle 1960s several writers, especially Blalock (1964) and Duncan (1966), argued that a more explicitly causal interpretation of results from partial correlation and regression procedures held considerable promise for the development of comprehensive models of social processes. An influential demonstration of the realization of that promise appeared shortly afterward (Blau and Duncan, 1967) and inspired a line of work loosely named "status attainment" research, which has retained its vigor. Mullins (1973, pp. 214–241) has described the extrasociological origins of causal modeling techniques and their extensive reorganization of the substantive interests of sociologists. A version of the triangulation doctrine mentioned above has also been incorporated by the causal modeling approach (Blalock, 1969a).

Two advantages of the approach were especially important in its early stages of growth in sociology. First, it provided a framework for formalizing many of the concepts and controversies that previously existed in the social stratification literature. For example, Featherman (1976) and Treiman (1976) discuss its importance in developing methods for determining structural and individual influences on occupational mobility, measuring trends in occupational mobility, and determining differences in the stratification processes that govern members of different subpopulations. Second, it provided a means of representing large multivariate systems in a concise manner (Blalock, 1969b). This made it easier for researchers to envision work on numerous specific topics as a more or less coordinated scholarly effort, even if that work did not always use the specific technical apparatus usually associated with causal modeling. Collins (1975, pp. 510–512) argues that competitive pressures increase the likelihood that members of a field will adopt techniques that make this sort of specialization possible.

Like other research techniques in the social sciences, causal modeling has no clear ties to substantive phenonema. Critics of the approach therefore characterize it as a "method in search of a substance" and portray resulting research as a mechanical enterprise with little theoretical inspiration (Coser, 1975). In contrast, adherents point to its role in providing a unified approach to the quantitative techniques of measurement and analysis of a variety of disciplines (for discussions of this role, see Mullins, 1973, pp. 235–237; Featherman, 1976, p. 24), and its continued use in a very large proportion of research in social stratification.

The inspiration for systematic empirical work on social stratification in science is usually credited to Merton's (1957) influential paper on scientific priority, but the empirical work did not start appearing until the mid 1960s. An important stimulus for the increased interest in such work was the adoption of causal modeling techniques by sociologists of science (Cole and Zuckerman, 1975, p. 157). Practicing sociologists of science made little or no contribution to the establishment of causal modeling techniques in sociology, but they were enthusiastic converts during the early stages of the spread of those techniques. An important event in the transmittal of the techniques to the sociology of science occurred at the sociology of science session during the 1966 annual convention of the American Sociological Association. A discussant, R. W. Hodge, one of the important figures in the dissemination of causal modeling techniques, used data from a paper presented by Cole and Cole (published in 1967) to show the superiority of causal modeling techniques over traditional contingency table analysis. The Coles subsequently published a monograph that employed the techniques extensively (1973).

Whether the interest in "the reward system of science" would have developed along the lines it did, or as quickly as it did, without the theoretical work of Merton and the empirical work of his students is moot; but one cannot help being struck by the formal similarities between the work in social stratification in the late 1960s and work on the reward system in science in the early 1970s. In one respect work on stratification processes among scientists was even more attractive than such work on the societal level. Thanks to the prior and independent development of the *Science Citation Index* (Garfield, 1964), it was possible to obtain measures of job performance for scientists, whereas this was a conspicuous omission from the status attainment models for the general population. Furthermore, the existence of easily accessible sources of biographical data on scientists (sources such as *American Men and Women of Science* and various disciplinary directories) and the development of more precise measures of institutional prestige (Cartter, 1966) made it easy for sociologists to enter the research area (the number of quantitative studies in the "sociology of sociology" began to grow prolifically at this time). Finally, in the United States at least, work on stratification in science could have been easily based on a fairly extensive and long-standing literature (Wilson, 1942; Caplow and McGee, 1958; Berelson, 1960) had Merton's theoretical work not been available.

To suggest the potential dispensability of a scholar's work does not depreciate the magnitude of his or her contribution (Merton, 1961). However, it does lessen the force of criticisms that hold Merton responsible for major features of research on the reward system in science. In particular, critics blame Merton for the logical positivist epistemology that they believe flaws the line of research (the links between Merton and logical positivism are sometimes greatly overdrawn; see, for example, Whitley, 1972), but they neglect the explicit logical positivism of the causal modeling approach itself (Blalock, 1964; Costner, 1969). Thus, although his celebrators may have overestimated Merton's role in the development of research on the reward system in science (Storer, 1973, pp. 281–285), his role is exaggerated to a much greater extent by his detractors.

Once established in the sociology of science, the causal modeling approach fostered lines of inquiry quite similar to those it prompted in the wider discipline. Since the techniques place a premium on the development of precise measuring instruments, their use has led naturally to critical appraisal of measures of important concepts and suggestions for their improvement (Hagstrom, 1971, pp. 379–380; Cole and Cole, 1973, pp. 21–36; Chubin and Moitra, 1975; Moravcsik and Murugesan, 1975; Lindsey and Staulcup, 1977; Porter, 1977). And, just as the causal modeling approach provides a coordinating framework for elaborating general models of stratification processes, its use in the sociology of science has led to the construction of more fully specific models of scientific reward systems (Allison and Stewart, 1975; Liebert, 1976; Long, 1977), analyses of differences between subpopulations within the system (Bayer and Astin, 1975; Reskin, 1976, 1978), and comparative analyses of reward systems in different disciplines and nations (Blume and Sinclair, 1973; Gaston, 1973, 1978). The potential for further elaboration along these and other lines remains substantial, and it therefore seems premature to conclude, as Nelson (1974) does, that work in this vein has spent itself.

Network Analysis and Invisible Colleges in Science

Unlike the methods of causal analysis, which were developed into a fairly coherent set of techniques before their adoption by sociologists of science, the various techniques of network analysis are not yet well consolidated. As a result, the use of particular network techniques in sociology and the sociology of science has been more tentative and sporadic than the adoption of causal modeling techniques. Furthermore, within the sociology of science, network techniques have played an ancillary, although crucial, role in work on already existing substantive issues, rather than providing models for importing new substantive concerns into the sociology of science. Thus, the impact of these techniques on the sociology of science has not been as great as the impact of causal modeling.

Network analysis can be traced to the work of Moreno (1934), who popularized the general approach under the term *sociometry*. Shortly after World War II, several scholars pointed out that sociograms could be represented as matrices of interpersonal choice and that various matrix manipulations could be employed to detect cliques, determine indirect connections, and derive measures of individual "status," or "centrality" (for general reviews of this work, see Abelson, 1967; Borgatta, 1968). During the 1950s graph theory also was employed as a means of representing sociometric data (Harary, 1959), and this representation led naturally to the extension of sociometric ideas to the analysis of transaction flows (Isard, 1960; Nystuen and Dacey, 1961). The transformation of sociomatrices into matrices containing measures of similarity of distance (MacRae, 1960) facilitated the application of techniques for the measurement of latent variables (at first, cluster and factor analysis; then, somewhat later, smallest-space analysis and multidimensional scaling techniques) to sociometric analysis. Sociologists played a major role in these developments. Such figures as Coleman (1964) and Goodman (1964) contributed important methodological work; and such areas as small-group research, urban network analysis, and labor mobility studies provided early examples of the use of resulting techniques.

Price's (1963) influential work on invisible colleges in science turned the attention of sociologists of science to problems of identifying such groups and determining their structural characteristics. Given the availability of network techniques, it is not surprising that sociologists of science started using them soon after the appearance of Price's work. Most of the early work examined either bibliometric ties (Garfield, Sher, and Torpie, 1964; Price, 1965) or informal communication patterns (Mullins, 1968; Crane, 1969; Crawford, 1971; Gaston, 1973).

Two major limitations of sociometric techniques as they existed during the middle 1960s had important effects on the nature of this work. First, in most instances the techniques could be applied only to groups of relatively small size (fewer than 200 members). In the case of informal communication networks, the interviews and questionnaires needed to provide complete information on a given relationship for all possible pairs of group members would become excessively long if more members were to be studied (the alternative of asking each respondent for a certain number of nominees does not provide complete information). In addition, even the largest computers could not perform the matrix operations involved in many of the techniques if the sociometric matrices were much larger than 200 by 200. Thus, even restricting attention to bibliometric structures did not enable investigators to transcend the size limitation. Because of this limitation, researchers were unable to use clique detection techniques for large groups of scientists (on the order of a discipline, for example) to determine whether they included the recognizable subgroupings suggested by the "invisible college" concept. Studies of the invisible college hypothesis were forced to employ alternate strategies, which yielded apparently inconsistent results. For example, Mullins (1968) used snowball sampling methods in a study of biological scientists and failed to find such subgroupings, while Crane (1969), using data on the internal sociometric structure of researchers studying the diffusion of agricultural innovations, concluded that the hypothesis was supported. In a subsequent series of papers, Griffith and Mullins (Griffith and Miller, 1970; Mullins, 1972; Griffith and Mullins, 1972) argued that the apparent inconsistencies could be explained by the hypothesis that the invisible college characteristics described by Price are present only during unique and relatively rare stages of the life cycle of a research area. Mullins and his colleagues (Mullins, 1973; Mullins and others, 1977) have presented further evidence in support of this claim.

The second major limitation of the network techniques that sociologists of science initially adopted was their inability to analyze several different sociometric ties simultaneously. Thus, it was possible to analyze a network of collaborations, a network of citation flows, and a network of informal communication, but not to analyze all three relationships jointly. This was an obvious weakness, since, in addition to the kinds of ties noted above, teacher-student relationships (Sturtevant, 1965; Ben-David and Collins,

1966; Fisher, 1967), joint participation in special conferences (Mullins, 1972), and a host of other ties (see Chubin, 1976) can be important features of the structure of research groups. In the face of this technical limitation, researchers often used information about one kind of tie to "validate" structural analyses based on another tie (for example, see Small, 1977) or simply analyzed the Boolean sum of the matrices of different ties to obtain a global structural picture (for example, see Crane, 1969, pp. 343–344). But neither of these two strategies allows the researcher to exploit all the structural information available in the different sociometric tie networks.

In the 1970s technical developments in network analysis enabled sociologists of science to avoid or overcome the two limitations discussed above. For example, Narin and his colleagues (Narin, Carpenter, and Berlt, 1972; Pinski and Narin, 1976) have studied citation relationships between scientific journals in an attempt to construct maps of science. By using journals rather than individuals or papers as the primary unit of analysis, they were able to study the structural characteristics of much broader sectors of science than is otherwise possible. But a high price is paid for this ability, since journals play an insignificant role in theories of the structural characteristics of scientific innovation. More recently, Small and Griffith have attempted to overcome the size limitations of existing sociometric techniques by analyzing co-citation ties; by examining networks of papers with high levels of co-citation, they generate structural representations of large sectors of science while retaining small-scale units of analysis (Small, 1973; Small and Griffith, 1974; Griffith and others 1974). Furthermore, because the extent to which a paper is co-cited with other papers can change from year to year, co-citation analysis—unlike other bibliometric relations, such as reciprocal citation or bibliometric coupling (Kessler, 1963)—provides a dynamic representation of current researchers' perceptions about interrelations among cited documents (Small, 1977). Thus, the analysis of co-citation ties may allow investigators to detect and study important cognitive changes in scientific development (Small and Stanley, 1976; Sullivan, White, and Barboni, 1977).

Co-citation analysis has an obvious advantage over methods

of identifying research areas of high interest that depend on scientists' perceptions of the past or bibliographical compilations, because the latter are often highly dependent on current interpretations of the nature of work in such areas (Kuhn, 1970, pp. 136–143). But co-citation analysis clearly does not constitute a panacea for all the difficulties involved in detecting and studying highly innovative groups of scientists. Because scientists' citation patterns involve both cognitive and social dimensions, co-citation ties reflect an amalgamation of these factors in unknown proportions. Early critiques of the use of co-citation analysis argued that authors of documents highly co-cited are not likely to constitute coherent social groups (Chubin, 1976), but evidence from the first few attempts to test the claim suggests that the authors do form such groups (Mullins and others, 1977; Small, 1977). Nevertheless, co-citation ties provide only a limited amount of structural information about the authors of highly co-cited documents. Studies of the structural characteristics of such groups therefore need to supplement co-citation data with data on other types of ties. It is also unlikely that studies based only on co-citation analysis can retrieve comprehensive lists of those who have contributed to the scientific developments they detect (Sullivan, White, and Barboni, 1977), and the effectiveness of co-citation analysis in fields with low rates of publication and citation—such as mathematics and the social sciences—is still open to question.

Recent years have also seen the development of techniques that permit the simultaneous analysis of networks consisting of several different types of ties (White, Boorman, and Breiger,1976; Burt, 1976). Once again, these technical innovations came from people outside the sociology of science (Small's work on co-citations constitutes the lone exception to this general rule). Rather than providing a solution to network analysis problems, however, these new techniques only dramatize the lack of a theoretical rationale for determining which particular ties are to be included in studies of the structural features of research groups (for a preliminary statement on this question, see Mullins, 1975) and, in the case of the technique suggested by Burt, how such ties might be weighted relative to one another. As a result, the first researchers to use these new methods (Breiger, 1976; Mullins and others, 1977) employed ad hoc and sometimes arbitrarily defined ties in their analyses. Considerably more work, involving both more precise theoretical specification and technical calibration, clearly will be required before research using these techniques can yield adequate structural representations of scientific research groups.

As noted earlier, new approaches to network analysis are still being developed, and the kind of interplay between technical innovation and substantive concern outlined above will probably also characterize future developments. Thus far, sociologists of science have used network techniques primarily for structural description rather than for detecting underlying principles that may govern network configurations (Holland and Leinhardt, 1975; Boorman and White, 1976). This is not a result of technical backwardness; it stems from the fact that substantive theory in the sociology of science does not yet embrace the latter issue. Like other methodological approaches in sociology, techniques for network analysis are only loosely related to substantive issues in the research areas that employ them.

Integration of Theory and Method

To claim that theory and method are integrated only very loosely in a field is to allow the possibility that particular techniques can inspire as much commitment among researchers as can certain theoretical insights (Kuhn, 1961, pp. 188–189; Masterman, 1970, p. 74). Of course, technique without clear theoretical grounding or theoretical insight without a technique for empirical exploitation is not likely to inspire the levels of commitment that can result where strong links between theory and methods exist. This lack of theory-method integration is probably a source of the pervasive eclecticism present in contemporary sociology and the sociology of science; it also explains why those who attempt to discredit research techniques by arguing that they are based on unreasonable theoretical premises (for example, see Law, 1974) do not convince those who use the techniques, and why those who call for the redirection of research effort along lines consistent with a new theoretical approach generally elicit a phlegmatic response unless such admonitions are accompanied by examples of appropriate techniques for such research. Current arguments for an interpretive sociology of science, or for incorporating the cognitive content of scientists' work in research in the sociology of science, rarely consider the question of appropriate techniques.

It is much easier to describe the gaps between theory and technique in the sociology of science than to predict the conditions under which the gaps might be narrowed (Kuhn, 1970, pp. 178–179). But one could argue that areas like the sociology of science —where commonality of substantive interest ensures that those with different theoretical and methodological orientations will maintain a relatively high degree of contact—enjoy a relatively high chance of experiencing a substantial narrowing of the gaps between theory and method. In contrast, in sociology as a whole, differences in theoretical and methodological orientations are likely to coincide with differences in substantive interest. This argument, which is based on a perspective that might be labeled "optimistic pluralism," is closely allied with arguments for methodological triangulation and takes heart from instances of fruitful interchanges between practitioners of different techniques. For example, early attempts to apply mathematical models in the sociology of science often lacked theoretical grounding (Gilbert and Woolgar, 1974, pp. 279–286), but recent efforts to establish links between the mathematical modeling and causal approaches (Allison and Stewart, 1974, 1975; Faia, 1975) hold promise for providing such grounding (Allison, 1976). Similarly, recent work on both quantitative and qualitative models of scientific citations (Allison, 1976, pp. 105–116, 160–166; Price, 1976; Gilbert, 1977) offer insights that may be consolidated to provide a more complete understanding of relationships between social recognition and social negotiation processes in science.

Hagstrom (1965, pp. 256–259) suggests that the "positivist ideal" is most prevalent in fields that enjoy high levels of consensus on methods and theories, but the ideal can also be present in the attenuated form characterized above as "optimistic pluralism" in fields where consensus is so low that intense dispute makes little sense to participants. In fields such as the sociology of science participants can maintain contact with colleagues of opposite opinions, and social solidarity can flow from the realization of understood disagreement (Scheff, 1967, pp. 39–46).

References

Abelson, R. P. "Mathematical Modes in Social Psychology." In L. Berkowitz (Ed.), *Advances in Experimental Social Psychology*. Vol 3. New York: Academic Press, 1967.

Allison, P. D. "Processes of Stratification in Science." Unpublished doctoral dissertation, University of Wisconsin, 1976.

Allison, P. D., and Stewart, J. A. "Productivity Differences Among Scientists: Evidence for Accumulative Advantage." *American Sociological Review,* 1974, *39,* 596–606.

Allison, P. D., and Stewart, J. A. "Reply to Faia." *American Sociological Review,* 1975, *40,* 829–830.

Bayer, A. E., and Astin, H. S. "Sex Differentials in the Academic Reward Structure." *Science,* 1975, *188,* 796–802.

Ben-David, J., and Collins, R. "Social Factors in the Origins of a New Science: The Case of Psychology." *American Sociological Review,* 1966, *31,* 451–465.

Berelson, B. *Graduate Education in the United States.* New York: McGraw-Hill, 1960.

Blalock, H. M., Jr. *Causal Inferences in Nonexperimental Research.* Chapel Hill: University of North Carolina Press, 1964.

Blalock, H., Jr. "Multiple Indicators and the Causal Approach to Measurement Error." *American Journal of Sociology,* 1969a, *75,* 264–272.

Blalock, H., Jr. *Theory Construction.* Englewood Cliffs, N.J.: Prentice-Hall, 1969b.

Blau, P. M., and Duncan, O. D. *The American Occupational Structure.* New York: Wiley, 1967.

Blume, S. S., and Sinclair, R. "Chemists in British Universities: A Study of the Reward System in Science." *American Sociological Review,* 1973, *38,* 126–138.

Boorman, S. A., and White, H. C., "Social Structure from Multiple Networks. II: Role Structures." *American Journal of Sociology,* 1976, *81,* 1384–1446.

Borgatta, E. F. "Sociometry." In D. Sills (Ed.), *International Encyclopedia of the Social Sciences.* Vol. 15. New York: Macmillan, 1968.

Breiger, R. L. "Career Attributes and Network Structure: A Block Model Study of a Biomedical Research Specialty." *American Sociological Review,* 1976, *41,* 117–135.

Burt, R. S. "Positions in Networks." *Social Forces,* 1976, *55,* 93–122.

Caplow, T., and McGee, R. *The Academic Marketplace.* New York: Basic Books, 1958.

Cartter, A. M. *An Assessment of Quality in Graduate Education.* Washington, D.C.: American Council on Education, 1966.

Chubin, D. "The Conceptualization of Scientific Specialties." *Sociological Quarterly,* 1976, *17,* 448–476.

Chubin, D., and Moitra, S. D. "Content Analysis of References: Adjunct or Alternative to Citation Counting?" *Social Studies of Science,* 1975, *5,* 423–441.

Cicourel, A. V. *Method and Measurement in Sociology.* New York: Free Press, 1964.

Cole, J. R., and Cole, S. *Social Stratification in Science.* Chicago: University of Chicago Press, 1973.

Cole, J. R., and Zuckerman, H. "The Emergence of a Scientific Specialty: The Self-Exemplifying Case of the Sociology of Science." In L.A. Coser (Ed.), *The Idea of Social Structure.* New York: Harcourt Brace Jovanovich, 1975.

Cole, S. "The Growth of Scientific Knowledge: Theories of Deviance as a Case Study." In L. Coser (Ed.), *The Idea of Social Structure.* New York: Harcourt Brace Jovanovich, 1975.

Cole, S., and Cole, J. R. "Scientific Output and Recognition: A Study of the Operation of the Reward System in Science." *American Sociological Review,* 1967, *32,* 377–390.

Coleman, J. S. *Introduction to Mathematical Sociology.* New York: Free Press, 1964.

Collins, R. *Conflict Sociology.* New York: Academic Press, 1975.

Coser, L. "Two Methods in Search of a Substance." *American Sociological Review,* 1975, *40,* 691–700.

Costner, H. L. "Theory, Deduction, and Rules of Correspondence." *American Journal of Sociology,* 1969, *75,* 245–263.

Crane, D. "Social Sturcture in a Group of Scientists: A Test of the 'Invisible College' Hypothesis." *American Sociological Review,* 1969, *34,* 335–352.

Crawford, S. "Informal Communication Among Scientists in Sleep Research." *Journal of the American Society for Information Science,* 1971, *22, 301*–310.

Duncan, O. D. "Path Analysis: Sociological Examples." *American Journal of Sociology,* 1966, *72,* 1–16.

Edge, D. "Why I Am Not a Co-Citationist." *Newsletter* of the Society for Social Studies of Science, 1977, *2,* 13–19.

Faia, M. A. "Productivity Among Scientists: A Replication and Elaboration." *American Sociological Review,* 1975, *40,* 825–829.

Featherman, D. L. "Coser's . . . 'In Search of Substance.'" *American Sociologist,* 1976, *11,* 21–27.

Fisher, C. S. "The Last Invariant Theorists: A Sociological Study of the Collective Biographies of Mathematical Specialists." *European Journal of Sociology,* 1967, *8,* 1094–1118.

Garfield, E. "*Science Citation Index*: A New Dimension in Indexing." *Science* 1964, *144,* 649–654.

Garfield, E., Sher, I. H., and Torpie, R. J. *The Use of Citation Data in Writing the History of Science.* Philadephia, Pa.: Institute for Scientific Information, 1964.

Gaston, J. *Originality and Competition in Science.* Chicago: University of Chicago Press, 1973.

Gaston, J. *The Reward System in British and American Science.* New York: Wiley-Interscience, 1978.

Gilbert, G. N. "Referencing as Persuasion." *Social Studies of Science,* 1977, *7,* 113–122.

Gilbert, G. N., and Woolgar, S. "The Quantitative Study of Science: An Examination of the Literature." *Science Studies,* 1974, *4,* 279–294.

Goodman, L. A. "A Short Computer Program for the Analysis of Transaction Flows." *Behavioral Science,* 1964, *9,* 176–186.

Griffith, B. C. and Miller, A. J. "Networks of Informal Communication Among Scientifically Productive Scientists." In C. E. Nelson and D. K. Pollock (Eds.), *Communication Among Scientists and Engineers.* Lexington, Mass.: Heath, 1970.

Griffith, B. C., and Mullins, N. C. "Coherent Social Groups in Scientific Change." *Science,* 1972, *177,* 959–964.

Griffith, B. C., and others. "The Structure of Scientific Literatures. II: Toward a Macro- and Microstructure for Science." *Science Studies,* 1974, *4,* 339–365.

Hagstrom, W. O. "Anomy in Scientific Communities." *Social Problems,* 1964, *12,* 186–195.

Hagstrom, W. O. *The Scientific Community.* New York: Basic Books, 1965.

Hagstrom, W. O. "Inputs, Outputs and the Prestige of University Science Departments," *Sociology of Education,* 1971, *44,* 375–397.

Harary, F. "Graphic Theoretic Methods in Management Sciences." *Management Science,* 1959, *5,* 387–403.

Hargens, L. L. *Patterns of Scientific Research: A Comparative Analysis of Research in Three Fields.* Washington, D.C.: American Sociological Association, 1975.

Holland, P. W., and Leinhardt, S. "Local Structure in Social Networks." in D. R. Heise (Ed.), *Sociological Methodology 1976.* San Francisco: Jossey-Bass, 1975.

Isard, W. *Methods of Regional Analysis: An Introduction to Regional Science.* New York: Wiley, 1960.

Kessler, M. M. "Bibliometric Coupling Between Scientific Papers." *American Documentation,* 1963, *14,* 10–25.

Knorr, K. D., "The Nature of Scientific Consensus and the Case of the Social Sciences." In K. D. Knorr, H. Strasser, and H. G. Zilian (Eds.), *Determinants and Controls of Scientific Development.* Hingham, Mass.: D. Reidel, 1975.

Kuhn, T. S. "The Function of Measurement in Modern Physical Science." *Isis,* 1961, *52,* 161–193.

Kuhn, T. S. *The Structure of Scientific Revolutions.* (2nd ed.) Chicago: University of Chicago Press, 1970.

Lammers, C. T. "Mono- and Poly-Paradigmatic Developments in Natural and Social Sciences." In R. Whitley (Ed.), *Social Processes of Scientific Development.* London: Routledge & Kegan Paul, 1974.

Law, J. "Theories and Methods in the Sociology of Science: An Interpretive Approach." *Social Science Information,* 1974, *13,* 163–172.

Liebert, R. J. "Productivity, Favor, and Grants Among Scholars." *American Journal of Sociology,* 1976, *82,* 664–673.

Lindsey, D., and Staulcup, H. "Improving Measures of Scientific Contributions to Knowledge." Paper presented at 72nd annual meeting of American Sociological Association, Chicago, Aug. 1977.

Long, J. S., "Productivity and Position in the Early Academic

Career: A Study of Two Cohorts of Ph.D. Biochemists." Unpublished doctoral dissertation, Cornell University, 1977.

MacRae, D. "Direct Factor Analysis of Sociometric Data." *Sociometry,* 1960, *23,* 360–371.

Masterman, M. "The Nature of a Paradigm." In I. Lakatos and A. Musgrave (Eds.), *Criticism and the Growth of Knowledge.* Cambridge, England: Cambridge University, Press, 1970.

Merton, R. K. "Priorities in Scientific Discovery: A Chapter in the Sociology of Science." *American Sociological Review,* 1957, *22,* 635–659.

Merton, R. K. "Singletons and Multiples in Scientific Discovery." *Proceedings* of the American Philosophical Society, 1961, *105,* 470–486.

Merton, R. K. "The Sociology of Science: An Episodic Memoir." In R. K. Merton and J. Gaston (Eds.), *The Sociology of Science in Europe.* Carbondale: Southern Illinois University Press, 1977.

Moravcsik, M. J., and Murugesan, P. "Some Results on the Function and Quality of Citations." *Social Studies of Science,* 1975, *5,* 86–92.

Moreno, J. L. *Who Shall Survive?* Beacon, N.Y.: Beacon House, 1934.

Mulkay, M. J. "Methodology in the Sociology of Science." *Social Sciences Information,* 1974, *13,* 107–119.

Mullins, N. C. "The Distribution of Social and Cultural Properties in Informal Communication Networks Among Biological Scientists." *American Sociological Review,* 1968, *33,* 786–797.

Mullins, N. C. "The Development of a Scientific Specialty: The Phage Group and the Origin of Molecular Biology." *Minerva,* 1972, *10,* 51–82.

Mullins, N. C. *Theories and Theory Groups in Contemporary American Sociology.* New York: Harper & Row, 1973.

Mullins, N. C. "A Sociological Theory of Scientific Revolution." In K. D. Knorr, H. Strasser, and H. G. Zilian (Eds.), *Determinants and Controls of Scientific Development.* Hingham, Mass.: D. Reidel, 1975.

Mullins, N. C., and others. "The Group Structure of Co-Citation Clusters: A Comparative Study." *American Sociological Review,* 1977, *42,* 552–562.

Narin, F., Carpenter, M., and Berlt, N. C. "Interrelationships of Scientific Journals." *Journal of the American Society for Information,* 1972, *23,* 323–331.

Nelson, B. "On the Shoulders of the Giants of the Comparative Historical Sociology of 'Science'—In Civilizational Perspective." In R. D. Whitley (Ed.), *Social Processes of Scientific Development.* London: Routledge & Kegan Paul, 1974.

Nystuen, J. D., and Dacey, M. F. "A Graph Theory Interpretation of Nodal Regions." *Papers and Proceedings of the Regional Science Association,* 1961, *7,* 29–42.

Pinski, G., and Narin, F. "Citation Influence for Journal Aggregates of Scientific Publications: Theory with Application to the Literature of Physics." *Information Processing and Management,* 1976, *12,* 297–312.

Porter, A. L. "Citation Analysis: Queries and Caveats." *Social Studies of Science,* 1977, *7,* 257–267.

Price, D. J. de S. *Little Science, Big Science.* New York: Columbia University Press, 1963.

Price, D. J. de S. "Networks of Scientific Papers." *Science,* 1965, *149,* 510–515.

Price, D. J. de S. "A General Theory of Bibliometric and Other Cumulative Advantage Processes." *Journal of the American Society for Information Science,* 1976, *27,* 292–306.

Reskin, B. F. "Sex Differences in Status Attainment in Science: The Case of the Postdoctoral Fellowship." *American Sociological Review,* 1976, *41,* 597–612.

Reskin, B. F. "Scientific Productivity, Sex, and Location in the Institution of Science." *American Journal of Sociology,* 1978, *83,* 1235–1243.

Scheff, T. J. "Toward a Sociological Model of Consensus." *American Sociological Review,* 1967, *32,* 32–46.

Small, H. G. "Co-Citation in the Scientific Literature: A New Measure of the Relationship Between Two Documents." *Journal of the American Society for Information Science,* 1973, *24,* 265–269.

Small, H. G. "A Co-Citation Model of a Scientific Specialty: A Longitudinal Study of Collagen Research." *Social Studies of Science,* 1977, *7,* 139–166.

Small, H. G., and Griffith, B. C. "The Structure of Scientific Literatures. I: Identifying and Graphing Specialties." *Science Studies,* 1974, *4,* 17–40.

Small, H. G., and Stanley, J. L. "Co-Citation Studies of Scientific Specialties: How Frequently Do Micro-Revolutions Occur in Science?" Paper presented at annual meeting of the History of Science Society, 1976.

Storer, N. W. "Prefatory Note to the Reward System in Science." In R. K. Merton, *The Sociology of Science: Theoretical and Empirical Investigations.* Chicago: University of Chicago Press, 1973.

Sturtevant, A. H. *A History of Genetics.* New York: Harper & Row, 1965.

Sullivan, D., White, D. H., and Barboni, E. J. "Co-Citation Analysis of Science: An Evaluation." *Social Studies of Science,* 1977, *7,* 223–240.

Treiman, D. J. "A Comment on Professor Lewis Coser's Presidential Address." *American Sociologist,* 1976, *11,* 27–33.

Webb, E. J., and others. *Unobtrusive Measures: Nonreactive Measures in the Social Sciences.* Chicago: Rand McNally, 1966.

White, H. C., Boorman, S. A., and Breiger, R. L. "Social Structure from Multiple Networks. I: Blockmodels of Roles and Positions." *American Journal of Sociology,* 1976, *81,* 730–780.

Whitley, R. D. "Black Boxism and the Sociology of Science: A Discussion of the Major Developments in the Field." In P. Halmos (Ed.), *The Sociology of Science.* Sociological Review Monograph No. 18. Keele, England: University of Keele, 1972.

Whitley, R. D. "Umbrella and Polytheistic Scientific Disciplines and Their Elites." *Social Studies of Science,* 1976, *6,* 471–497.

Whitley, R. D. "The Sociology of Scientific Work and the History of Scientific Developments." In S. S. Blume (Ed.), *Perspectives in the Sociology of Science.* New York: Wiley, 1977.

Wilson, L. *The Academic Man.* New York: Oxford University Press, 1942.

Zelditch, M. "Some Methodological Problems of Field Studies." *American Journal of Sociology,* 1962, *67,* 566–576.

7

Donald deB. Beaver

Possible Relationships Between the History and Sociology of Science

Paraphrasing Kant, one philosopher of science has said, "Philosophy of science without history of science is empty; history of science without philosophy of science is blind" (Lakatos, 1971, p. 91). To judge from recent criticisms, an appropriate parallel for the history and sociology of science might be, "The history of science without the sociology of science is myopic; the sociology of science without the history of science is superficial." Many individuals in each discipline, however, perceive neither need nor special value in interrelationships. In view of the relative youth and vigor of the two fields, and their great variety of aims and methods, such independence may not be a handicap. But such perceptions are not necessarily accurate or desirable, for each discipline has much to offer the other; and the respective situations of historians and sociologists of science might be improved if they expanded their concerns to include an appreciation of each other's methods, goals, and knowledge.

This chapter contains fundamental observations about the nature of the sociology and the history of science and the relationships that may be realized in the future. (I shall not discuss current research areas in the field; readers unfamiliar with them may find excellent descriptions elsewhere in work that, to some degree, anticipates this essay [see MacLeod, 1977; Thackray, 1978].) A comprehensive literature exists in the history and the sociology of science describing the origins, growth, research areas, orientation, methods, goals, relationships, accomplishments, present analyses, current activities, and suggestions for future directions.[1]

Any discussion must limit itself to what is possible, given demands for space, and must recognize that current possibilities for cooperation may change. The subject—science, scientists, and society—is influenced by dynamic processes that will affect the research and relationships in the future. Big Science, the primary focus of contemporary sociology of science, is experiencing novel changes as funding limits appear to be reached. Consequently, intellectual migration from current interests will likely create new research strategies that appear promising, given the competition for social and intellectual resources (see Mulkay, 1972). This potential for change is also true for the history and sociology of science as they face problems similar to those sciences they study. Finally, the problems that scholars choose to define and investigate will naturally affect the relationships between these specialties.

Although there is a long tradition of scholarship in the history of science, its emergence as an established discipline with a sizable body of specialized practitioners is virtually contemporaneous with that of the sociology of science. Both disciplines grew and prospered along with Big Science following World War II, after having undergone a period of latency and limited development. In the United States, each field has an acknowledged founder: George Sarton, whose American work began in 1915, and Robert K. Merton, whose studies in the (historical) sociology of science began in the 1930s. The interaction of these two men at Harvard represents one of the earliest relationships between the two disciplines, as does Merton's doctoral thesis, which has been seen both as historical sociology of science and as social history of science (Merton, 1938). Institutionally, the history of science is the older field. Its texts and professional academic appointments (1920s), the History of Science Society (1924), and its journal, *Isis* (1913), predate their counterparts in the sociology of science, whose society was founded in 1975 and for which no journal has been established.

The history of science has passed through a number of changes in emphasis, beginning with litanies of scientific heroes, an associated quest for heroic precursors, and a moralistic spirit of scientific humanism. Currently, the established profession includes a large variety of methodological and historiographical approaches, ranging from the reconstruction of ancient Babylonian astronomy to the oral history of twentieth-century science. The clearest division in the field is that between the "externalists" and the "internalists" (Basalla, 1968, 1975; Kuhn, 1968; MacLeod, 1977). Fortunately, debate between these positions appears to be subsiding, and new studies fitting neither category have developed. Nonetheless, the terms will serve here to symbolize important differences of outlook in the field, even though it may be more accurate to recognize a continuum of approaches.

Whereas the internalists are concerned primarily with the genealogy of scientific ideas—that is, with the logic and content of their genesis, technical development, and reception—the externalists are interested mainly in the social factors pertaining to scientific activity, such as those involved in organizations (laboratories, societies, universities), popular attitudes toward science, the interaction of science and society, and the diffusion of technical knowledge. Put in the extreme, the content of science is primary for the internalist and secondary for the externalist historian of science: and the social context of science is primary for the externalist and secondary for the internalist. Within the externalist camp, there is further division along qualitative and quantitative methodological lines; the latter is developing increasing connections with the sociology of science and scientometrics (Thackray, 1978).

A parallel to the internal-external division, although not quite so pronounced, exists in the sociology of science, along the "cognitive" and "social" dimensions. Those concerned with the sociology of knowledge, "externalists" in the sociology of science, play a role similar to that of their internalist counterparts in the history of science (Whitley, 1972; MacLeod, 1977; Thackray, 1978).

If generally more restricted than the history of science with respect to the time periods of the scientific activity it investigates, the sociology of science has seemingly greater diversity in its approaches and foci of interest in the social aspects of science. There is no satisfactory comprehensive taxonomy of research areas to help define and describe the discipline, most probably because—apart from being either present or past oriented, qualitative or quantitative—studies may categorize their subject matter in well over a half dozen ways (Storer, 1966, chap. 1, 1973; Ben-David, 1971, chap. 1; Whitley, 1972; Blume, 1974, chap. 1; Cole and Zuckerman, 1975; Merton, 1977). However, over and above the possibility of over two dozen different types of sociological studies of science, two perspectives are convenient and appropriate for briefly summarizing activity in the field: the institutional and the interactional (Ben-David, 1971). In the institutional perspective, attention centers on the interaction and mutual influence of social organization and cognitive structures; the social processes of scientific innovation; and the emergence, growth, and development of scientific specialties. Interactional studies are more concerned with the social activity of science as a communication system. They focus on research organization and productivity, systems of reward and evaluation, patterns of communication, and the norms and values of science. Still other prominent research areas involve relationships between scientific institutions and the larger society; factors influencing the popularization of scientific knowledge; and, most recently, historical and sociological prosopography (Shapin and Thackray, 1974; Merton, 1977; Pyenson, 1977).

In broader perspective, the sociology of science belongs with the history, psychology, and philosophy of science in the social studies of science, as distinguished from science policy studies, a derivative of political science. Between those two lies scientometrics, which concentrates on the statistical study of scientific activity (Spiegel-Rösing, 1977).

Certain research interests in the history of science appear virtually indistinguishable from those in the sociology of science. As early as 1963, one sociologist of science commented that some of "the newer generation of historians of science . . . produce works . . . which are so nearly oriented to the explicit theoretical concerns of the sociologist that only the least effort is required to show how they exemplify or develop those sociological concerns" (Barber, cited in Merton, 1977, p. 72). MacLeod (1977, p. 161) notes that studies in the social history of science now emphasize "scientific institutions (including societies and universities); the scientific professions; scientific disciplines, specialties, and research programs; and science in relation to wider social developments." Nonetheless, despite similarities, certain marked differences between the history and sociology of science must be recognized and allowed for in any discussion of possible interrelationships.

Constraints on Relationships

Although the differences between the two disciplines constitute barriers to natural interaction, they simultaneously promise that hybrid vigor may be expected from crossing separate varieties of the science studies species. Three of these differences—orientation to time, self-image, and basic units of analysis—deserve attention, because they substantially affect the nature and types of possible relationships.

Time. Most work in the sociology of science focuses on relatively recent scientific activity. In the beginning, the field was far more historically oriented and thus more difficult to distinguish from the social history of science (Merton, 1938; Bernal, 1939; Barber, 1952). With a few exceptions (for example, Kuhn, 1962; Ben-David, 1971; Mulkay, 1972), much stronger differences have subsequently appeared (contrast Cardwell, 1957; Mendelsohn, 1964; Crosland, 1967; Gilpin, 1968; Hahn 1971, with Storer, 1966; Crane, 1972; Cole and Cole, 1973; Gaston, 1973; Zuckerman, 1977). Merton's earlier work, for example, contains abundant material from the more remote past of science, but his later work contains relatively less (Merton, 1973). The discipline's emphasis on recency has grown along with its increased cohesion and identity, so that, viewed in terms of temporality, the sociology of science is largely a transverse but not a longitudinal enterprise (Cole and Zuckerman, 1975).

The history of science, in contrast, contains relatively little work on the recent past. Few, if any, professional studies have ventured much closer to the present than the end of World War II.

For example, no published history of mathematics or biology deals in comprehensive fashion with twentieth-century developments—largely because the cumulation, specialization, and exponential growth of science make a full treatment of recent science virtually impossible without teamwork, itself an unlikely research style in a humanistically oriented field. Moreover, quite apart from whatever intrinsic difficulties exist in writing the history of recent science, historians in general are wary of treating the immediate past.

In short, as the disciplines are presently constituted, a fundamental difference in orientation to time separates the history and sociology of science. As that separation has grown, it has increasingly restricted the possibility of interrelationships.

Self-Image. The two disciplines also differ in the way they conceive of their place in the spectrum of professional scholarship. Most historians of science, like most historians in general, consider their discipline a part of the humanities, as opposed to the social sciences. Their orientation reflects a strong emphasis on the philosophical, history-of-ideas approach especially associated with the scholarship and influence of Alexandre Koyré. Some, however (invariably externalists, especially those who are quantitatively oriented), consider that their work falls more naturally with the social sciences. In contrast, sociologists of science generally agree that their field constitutes a social science; comparatively few of them—those who are inclined toward philosophy, the history of ideas, and the sociology of knowledge—have concerns closer to the humanities.

In two ways the literature of the history of science resembles that of the humanities much more than that of the social sciences. In the history of science, the time distribution of citations to past literature is approximately uniform; that is, unlike the situation that obtains in the sciences, there appears to be no preference for the present over the past.[2] Its citation patterns lack that characteristic "immediacy" phenomenon of the natural sciences and some social sciences, which use recent work much more than older work (Price, 1970). Instead, the growth curve of the cumulated number of citations over time appears to be linear, which parallels what one would expect if the history of science concentrated on the major ideas and events of past science, for that is precisely the characteristic displayed by the growth of eponymic items in science (Beaver, 1976). In contrast, the literature of the recent past of the sociology of science displays patterns similar to those found in the consolidation of an active scientific research front: a monotonically decreasing average age of citations and the presence of an immediacy factor comparable to "rapidly growing specialties in the physical sciences, and much higher than that for sociology as a whole" (Cole and Zuckerman, 1975, p. 150).

An additional indicator of opposite self-conceptions is provided by the differential practice of collaborative research. The spread of teamwork as a typical research style and strategy attendant on the professionalization of science (Beaver and Rosen, 1978) has begun to be paralleled in the social sciences. Currently between one fifth to half of the articles published in economics, psychology, and sociology had multiple authors. For sociology as a whole, the growth of collaborative research has reached the point where, for 1966–1975, it accounts for 37 percent of the articles published in core journals. In the sociology of science, 20 percent of the research articles of the past twenty years have been collaborative. That incidence has also increased with time; for the most recently measured period, 1966–1973, it reached 31 percent (Cole and Zuckerman, 1975). In the history of science, there is no sign that multiple authorship is increasing or that it represents an attractive research style. In the core journal, *Isis,* the percentage of collaboratively authored research has remained quite constant over the past twenty or so years, at about 5 percent (21 of 405 papers, 1955–1976). The situation is quite similar for *History of Science,* where the percentage is 2.6 (3 of 114 papers, 1962–1976), but it changes in the expected direction for *Social Studies of Science,* where 20.8 percent of the articles are multiply authored (16 of 77 papers, 1971–1977).

Thus, in addition to qualitative judgments, quantitative measures of patterns in citation and in collaborative research over time operationally confirm the divergent research styles and orientations of the two fields. During the recent past, in which the two disciplines have come of age, the literary artifacts of their research styles have exhibited significant and growing differences.

Such divergence is not due to the subjects. Relationships appear to be closer and less differentiated in Western Europe. And where history is conceived of as a science, in the Soviet Union, for example, the two fields can be seen as aspects of *naukovedenie,* (social studies of science) and sociological studies have been supported by institutes in the history and in the philosophy of science. It is not surprising in the USSR to find a historical perspective in the sociology of science because the social history of science played a major role in the revival of *naukovedenie* in the 1960s (Lubrano, 1976; Merton and Gaston, 1977).

Units of Analysis. Potential interaction is also affected by differences in the preferred objects of study: for the historian, the individual; for the sociologist, the group. Some lines of research in the spectrum of historical scholarship tend toward a group focus, but in general the historian prefers to concentrate on individual ideas or scientists and the relationships among them, whereas the sociologist prefers collectivities. When concerns about group behavior enter historians' accounts, they do so most often in terms of generalizations serving as contextual or background material. Similarly, the individual appears in sociology most frequently through interview protocols used to collect data to exemplify and interpret generalizations. That difference in outlook is expressed in a recent book review: "They have in effect written a sociologically oriented history of the field but I suspect that a similar study by a historian of science would have focused upon a particular person, group, event, or problem area" (Crane, 1977, p. 27). (The same review seems also to imply that theoretical articulation is more important to the sociology than to the history of science, which may reflect yet another aspect of their differing scholarly orientations.)

Sociology of science aims at statistical generalizations, truths that by nature are like the statistically based principles of modern science, in contrast to the more particularistic and individualistic findings of the history of science. The two fields are complementary, each illuminating different aspects of science: science as knowledge and science as social activity. Since each, by nature of its focus and purposes, tends to exclude the other's truths, interaction is most likely to consist of borrowing useful results. From this point of view, complementarity also applies to the internalist and externalist history of science: "Although the internal and external approaches to the history of science have a sort of natural autonomy, they are in fact, complementary concerns" (Kuhn, 1968, p. 81). The sociologists and the externalist historians of science therefore have much in common.

Possible Relations

Recognition of these significant differences between the history and sociology of science considerably simplifies discussion of their potential interrelationships. The different kinds of possible relationships with the sociology of science depend on whether they involve the external or the internal history of science. Joint research is possible with the externalist and the sociologist of science but unlikely with the internalist historian and the sociologist. Interactions along a variety of research lines involving external historians and sociologists are already well on the way to institutionalization with the new Society for the Social Studies of Science (4S), its *Newsletter,* and such journals as *Social Studies of Science* and the forthcoming *Scientometrics.* The externalists' joining with sociologists appears to be the normal consequence of the general history and sociology of science having grown further apart in their concerns and having reached a size that characteristically sees specialties splitting into divergent research programs (Price, 1963). So it was fortuitous that sociologists were prepared to start a new society at the time that externalist historians were reaching the point of regrouping.

For any historian of science, the most obvious benefit of a more direct familiarity with the sociology of science is the acquisition of new perspectives that might generate new historical insights and suggest new problem areas. For the sociologist, historical data may introduce puzzles and new lines of research, increase the range and accuracy of concepts, and refine and extend sociological theories. The injection of historical dimensions into the sociology of science might help it clarify issues as Kuhn's (1962) *Structure of Scientific Revolutions,* a sociologically oriented work, helped the history of science to clarify some of its issues. The reactions of historians of science to Kuhn's book—their resistance to and debates about his work—sharpened historians' self-consciousness and provided models for new approaches in investigation and analysis.

The quantitative approach, a rapidly growing area in the

externalist history and sociology of science, permits the increasing use of the research tool known as *prosopography,* an occasion for greater possibilities of historical-sociological interaction, especially in the study of "low" as opposed to "high" science and in the study of popular attitudes toward science (Shapin and Thackray, 1974; Pyenson, 1977).

The latter area appears especially promising, for it deals with the problems of how the general public becomes aware of scientific developments; what its attitudes to science and scientific activity are; how these originate, develop, and are influenced; and what consequences public attitudes toward science have for its development. Such questions have been relatively neglected in the history of science—partly because of the strong appeal of the dominant internalist tradition and partly because of the difficulty of acquiring and processing information. Much of what has been done has concerned the history of scientific societies; the motivations, characteristics, and attitudes of society members and supporters; and the relations of scientific societies with the surrounding community (Ornstein, [1913] 1938; Schofield, 1963; Crosland, 1967; Purver, 1967; Hahn, 1971; Webster, 1975). More recent studies are demonstrating greater concern for popular attitudes (Sinclair, 1974; Thackray, 1974; Kohlstedt, 1975; Oleson and Brown, 1976).

Because of the difficulties involved in conceptualization and in obtaining data, the study of popular attitudes is particularly challenging. It is also essential to an understanding of the social context in which scientific activity occurs; its greatest potential value would be realized in comparative studies across both cultures and scientific fields. Thus, history might aid sociology of science in formulating theoretical models for understanding the diffusion and incorporation of scientific knowledge in the world at large. In turn sociologists' theoretical understanding may help the historian of science assess the role that such social attitudes have played in influencing scientists' careers.

In certain respects, then, sociologists and externalist historians of science can profitably interact. In addition, the sociology of science also could benefit from increased attention to the internal history of science. Internalist historians of science, as well as sociologists of science see a dismaying propensity for the sociology of science to cast its problems in such a way as to avoid the content of science in any direct confrontation. They therefore claim that its results are empty; for, without content, nothing validates the word *science* in the sociology of science; its analyses might apply to anything at all. "Any attempt at a sociological understanding of science must, in my view, involve an understanding of what it is that scientists produce" (Whitley, 1972, p. 63). "Sociologists, in their attempt to establish an independent sub-discipline in the Sociology of Science, have too often disregarded work in the History and Philosophy of Science" (Whitley, 1972, p. 61). Similarly, in criticizing various sociological models of scientific development, Parker (1975) suggests that what they lack is consideration of the distinctive nature of scientific knowledge. A more distant observer reflected summarily that, "Recently, it should be noted, there has been renewed observation that the nature and direction of scientific growth cannot be adequately understood without dealing specifically with the contents of science—its concepts, data, paradigms, and methods. The idea that the development of science can be analyzed at all effectively, apart from the concrete research of scientists is said to have proven false" (Whitley, 1972; Storer, 1973, p. xvii [with references to Mulkay, 1969; Barnes and Dolby, 1970; King, 1971]; Parker, 1975).

Where the emphasis remains on the present in the sociology of science, the history of science will probably play only a small role in supplying the desired content. But where the emphasis is on process rather than structure, where concern lies with the dynamic development of science, then both the viewpoint and the content of the history of science become valuable. Just as general sociology must confront the historical (Berger, 1963), so the sociology of science must cultivate the history of science in order to see the scientific world as it truly is. That is, the sociologist must understand something of the traditions from which have developed the aims and aspirations of the scientists whose behavior is under investigation. Both the history and the sociology of science share a common skepticism, they do not trust their subjects to report faithfully and fully on what they do. Unlike sociologists of science, the internalist historian of science is interested in scientific content, in distinguishing between what scientists say that they do and think and what in fact occurs (Brush, 1974). Thus, internalist history of science can provide an experienced corrective vision regarding the difference between the actual and the reported development of scientific knowledge and activity; it represents a valuable resource for putting scientists' statements in perspective. Still further value results from its ability to aid in discerning what is recent from what is traditional and in providing a historical testing ground for the accuracy and validity of the sociologist's insights.

Some of the appeals to include the content of science in the sociology of science reflect the view that emphasizing its logical and theoretical dimensions will suffice. The internal history of science could help expand that limited view by indicating that the research equipment necessary for creating and testing scientific knowledge is an integral part of its substance (Kuhn, 1962; Ravetz, 1971). That is, the sociology of science requires a theory of scientific knowledge that incorporates instrumentality as well as abstract thought.

One of the major themes in the history of the emergence of modern science is that it involved the fusion of craft and scholarly traditions, so that *knowing* could no longer be separated from *doing*. Yet the innovative role of the craft tradition in the production of knowledge and new paradigms is too often overlooked. That marriage of hand and mind requires attention to the empirical and instrumental sides of science, to what scientists do as well as what they think, and how what they do and think influences their social behavior. Despite emphasis in the history of science on the new instruments of the scientific revolution, insufficient attention has been given to the way that these techniques affect scientific activity and scientists' social behavior. A recent study represents a notable exception, in that it pays attention to the relationships among technology, social behavior, and cognitive structures, finding that various research equipment influenced the social composition of research groups and their research strategies (Edge and Mulkay, 1976). Although experimental-theoretical dimensions have already proved useful in some sociological studies of science (Hagstrom, 1965; Gaston, 1973), they have not generally been employed with respect to the underlying character of scientific knowledge.

Finally, the peculiar content of a scientific discipline affects the social behavior of its research scientists and also affects its cognitive structures (Menard, 1971). In the construction of general sociological theories, a significant problem consists in identifying and evaluating the biases introduced by the choice of disciplines or research problem area chosen for investigation. Just as philosophies of science vary in accordance with the sciences chosen as source and exemplification, sociologists' choices of different scientific fields may lead to different conceptions and emphases. Thus, a potentially valuable contribution to the sociology of science in conceptual development and the testing of theory might result from internalists' participating with sociologists in comparative historical-sociological studies of scientific fields.

In short, the sociology of science might benefit from increased attention to the internalist history of science in at least three general ways. Historians can (1) aid in distinguishing reality from myth in scientists' self-referential statements, as it also provides a historical laboratory for sociological investigation and analysis; (2) articulate, illustrate, and emphasize the essential role of the technology of research instruments and experiment in the production and definition of scientific knowledge; and (3) provide a clearer idea of how the subject matter of scientific fields influences social behavior.

If the history of science contributes to sociology through an injection of positivism, because its contributions derive from a close study of scientists' commitments to an ordered, objective, external reality and to facts independent of social consensus, sociology of science contributes to the internalist history of science through an injection of relativism. Internalists are willing to admit that social factors may influence the pace and direction, but not the content, of scientific research which is their primary concern. So sociology's influence will be minimal. The probability of contributions to internalist history derives from sociology's statistical nature increasing with the recency of historical period. That potential contribution is nonexistent for Babylonian astronomy; it is more probable for twentieth-century physics.

The statistical methods of sociologists may offer some guidance to the internalist history of science. For example, the internalist historian of science is frequently interested in whether or not scientists act in traditional or novel ways, not only intellectually but also in interactions with peers. The historian's interest in social interaction is similar to the sociologist's interest, but, because most internalists focus on individuals, their generalizations in this area are limited and particularistic. All scientists probably know the current social conventions for printing names on multiply authored papers, but what they "know" is usually limited to their own specialty. Scientists often believe that every specialty does it the same way, but that is not correct according to studies conducted by sociologists of science (Zuckerman, 1969). Just as there appear to be as many histories as there are historians, the norms of name ordering yields as many traditions as there are fields and subfields of research. Internalist historians could invest their history with context on such matters as name ordering, informed by the sociology of science, if it were not the case that sociologists of science generally lack historical awareness and concern with scientific content.

Team research is another potentially fruitful area for interaction between historians and sociologists of science, because its growth spans the continuum of scientific creativity from individual to group. The individual (the traditional unit of analysis for the historian of science) is gradually being replaced by the group (the focus of sociological inquiry). Sociological understanding of the processes of group interactions could help illuminate historical questions. Theories and models from the sociology of science are important to historians seeking to understand the historical production of scientific knowledge. Reciprocally, as historial research into processes of interaction in groups of scientists accumulates, internalist historical work (on laboratory notebooks, diaries, correspondence, and the like) when available in sufficient quantities may provide the possibility for statistical analysis by sociologists.

Finally, the mind of the scientist has often proven frustratingly elusive for the historian of science engaged in reconstructing the development of scientific ideas and the creation of scientific knowledge. As Albert Einstein (1949, p. 23) noted, "The essential in the being of a man of my type lies precisely in *what* he thinks and *how* he thinks, not in what he does or suffers." In contemporary group research, necessitated by Big Science, what and how scientists think—including their attitudes and approaches to research strategy and tactics, and their criteria for the evaluation of the adequacy and validity of others' research—must now become easier to study because the process of thinking about research is done more openly, in a group session, unlike the solitary work of an Einstein. Insofar as collaborative research requires the logic of scientific inquiry to become public, speculation about the hitherto private world of the scientist's mind may become more subject to scientific investigation by historians and sociologists.

The Realization of Cooperative Relationships

Aside from current relationships now maintained by sociologists and historians, future relationships will probably result from the individual idiosyncratic behavior of scholars. The role of chance and serendipity in the career of Thomas Kuhn, as perceived by Merton (1977), is an example of how scholars are influenced by other fields, a process which often produces innovative ideas.

Relationships between sociologists and historians will be inhibited because of the way research specialists define their fields. They tend to define certain things as "within the specialty" and other things as outside. Thus, relationships between fields depend on the types of questions individuals define and wish to investigate. Questions that researchers ask which involve areas of knowledge outside their own field do not conform to the usual questions studied, or at least they are not central to their own field; the questions are more difficult to evaluate or to pursue than questions from within one's discipline. Unless a new discipline requiring training in both history and sociology of science emerges, barriers to interrelationships are severe and virtually require that relationships be developed by the "occasional odd scientist or scholar" (Merton, 1977, p. 28; see also Price, 1965).[3]

In a collaboration between a sociologist and a historian of science, each will have interests, concerns, and conceptions of minor interest to the other. Each will want to pursue certain avenues of investigation that the other sees as trivial, unimportant, or simply misleading—leading, that is, away from the real problem and its satisfactory solution. Even though the central issues may be essentially the same, they will not be perceived to be so and will be analyzed and conceptualized in different fashions, so that part of the working relationship requires reciprocal education.

Experience attests to the value of the informed sociological imagination in aiding attempts to decipher the meaning of social patterns in the history of science. Conversely, historical information, which the sociological imagination must have in order to function helpfully and pertinently, benefits the sociology of science as the information helps to evaluate the applicability and validity of different theoretical models.

From the perspectives of the parent disciplines, something valuable is lost in such a situation because what might otherwise have received a fuller and more detailed treatment has to be compromised to permit a mutually satisfactory definition of the problem and its resolution. Yet something valuable is also gained in the production of knowledge, which illuminates in new ways both disciplines rather than just one, however problematic it may be to locate the new perspectives in each discipline's cognitive structure.

Conclusion

Social historians and historical sociologists of science would benefit substantially from increased relationships between their two fields. Similarly, sociologists of science with interests in the sociology of knowledge stand to gain from increased attention to internalist history of science and philosophy of science, but the exchange is asymmetrical. Although internalist historians of science have some interest in the social context of the development of scientific ideas and scientists' behaviors, they believe that the sociology of science can do little to help promote a deeper understanding of what they wish to reconstruct.

Excluding the interdisciplinary efforts in social studies of science, the possible interrelationships of the history and sociology of science will depend on the research interests of individual scholars. The future development of such interrelationships depends on how well their results fulfill their aims and contribute to the disciplines. If superior history of science and sociology of science result from attention to each other's disciplines, then interactions will multiply.

Notes

1. For overviews of the two fields, see Barber, 1968; Kuhn, 1968; Spiegel-Rösing, 1973. For the past development and current state of theoretical commitments in the sociology of science, see Storer, 1973; Cole and Zuckerman, 1975; Merton, 1977. For developments outside the United States, see Lubrano, 1976; Merton and Gaston, 1977. For the social history of science, see MacLeod, 1977; Thackray, 1978. For the history of science, see Thackray and Merton, 1972; Multhauf, 1975; Thackray, 1975; Basalla, 1975; Price, 1975. For its major historiographic statements, see Kuhn, 1962; Agassi, 1963. Critical bibliographies for the history of science appear annually in *Isis;* see also the *Isis Cumulative Bibliography* (1971). For a recent bibliography of the sociology of science, see Spiegel-Rösing, 1973.

2. This observation is tentative. My sample size was small, and up to 1945 the data were references to primary sources; after 1945, the references were to history of science papers. Even with this methodology, no turning points were found. A more detailed analysis, if results were similar, would lend support to my observation.

3. It would be instructive to perform a citation analysis of the history and sociology of science. Cole and Zuckerman (1975) did that for the sociology of science, and the history of science is notable for its lack of representation. Personal impressions from the journal literature of the history of science are that sociology of science appears perhaps even less in the history of science.

References

Agassi, J. *Towards an Historiography of Science,* The Hague, Netherlands: Mouton, 1963.

Barber, B. *Science and the Social Order.* New York: Macmillan, 1952.

Barber, B. "Review of T. S. Kuhn, *The Structure of Scientific Revolutions.*" *American Sociological Review,* 1963, *28,* 298–299.

Barber, B. "The Sociology of Science." In D. L. Sills (Ed.), *International Encyclopedia of the Social Sciences.* Vol. 14. London: Macmillan and Free Press, 1968.

Barnes, S. B., and Dolby, R. G. A. "The Scientific Ethos: A Deviant Viewpoint." *European Journal of Sociology,* 1970, *11,* 3–25.

Basalla, G. (Ed.). *The Rise of Modern Science: Internal or External Factors?* Lexington, Mass.: Heath, 1968.

Basalla, G. "Observations on the Present Status of History of Science in the United States." *Isis,* 1975, *66,* 467–470.

Beaver, D. deB. "Reflections on the Natural History of Eponymy and Scientific Law." *Social Studies of Science,* 1976, *6,* 89–98.

Beaver, D. deB., and Rosen, R. "Studies in Scientific Collaboration, I." (Two-part continuation in press.) *Scientometrics,* 1978, *1,* 65–84.

Ben-David, J. *The Scientist's Role in Society.* Englewood Cliffs, N.J.: Prentice-Hall, 1971.

Berger, P. L. *Invitation to Sociology: A Humanistic Perspective.* New York: Doubleday, 1963.

Bernal, J. D. *The Social Function of Science.* London: Routledge & Kegan Paul, 1939.

Blume, S. S. *Toward a Political Sociology of Science.* New York: Free Press, 1974.

Brush, S. "Should the History of Science Be Rated X?" *Science,* 1974, *183,* 1164–1172.

Cardwell, D. S. L. *The Organization of Science in England.* London: Heinemann, 1957.

Cole, J.R., and Cole, S. *Social Stratification in Science.* Chicago: University of Chicago Press, 1973.

Cole. J. R., and Zuckerman, H. "The Emergence of a Scientific Specialty: The Self-Exemplifying Case of the Sociology of Science." In L. A. Coser (Ed.), *The Idea of Social Structure.* New York: Harcourt Brace Jovanovich, 1975.

Crane, D. *Invisible Colleges: Diffusion of Scientific Knowledge in Scientific Communities.* Chicago: University of Chicago Press, 1972.

Crane, D. "Review of D. O. Edge and M. J. Mulkay, *Astronomy Transformed: The Emergence of Radio Astronomy in Britain.*" In Society for Social Studies of Science *Newsletter,* 1977, *2* (4): 27–29.

Crosland, M. *The Society of Arceuil: A View of French Science at the Time of Napoleon I.* London: Heinemann, 1967.

Edge, D. O., and Mulkay, M. J. *Astronomy Transformed: The Emergence of Radio Astronomy in Britain.* New York: Wiley-Interscience, 1976.

Einstein, A. "Autobiographical Notes." (P. A. Schilpp. Trans.) In P. A. Schilpp (Ed.), *Albert Einstein: Philosopher-Scientist.* New York: Tudor, 1949.

Gaston, J. *Originality and Competition in Science.* Chicago: University of Chicago Press, 1973.

Gilpin, R. *France in the Age of the Scientific State.* Princeton, N.J.: Princeton University Press, 1968.

Hagstrom, W. O. *The Scientific Community.* New York: Basic Books, 1965.

Hahn, R. *The Anatomy of a Scientific Institution: The Paris Academy of Sciences, 1666–1903.* Berkeley: University of California Press, 1971.

Isis Cumulative Bibliography, 1913–1965. 2 vols. London: Mansell, 1971.

King, M. D. "Reason, Tradition, and the Progressiveness of Science." *History and Theory,* 1971, *10,* 3–32.

Kohlstedt, S. *The Formation of the American Scientific Community.* Urbana: University of Illinois Press, 1975.

Kuhn, T. S., *The Structure of Scientific Revolutions.* Chicago: University of Chicago Press, 1962.

Kuhn, T. S. "The History of Science." In D. L. Sills (Ed.), *International Encyclopedia of the Social Sciences.* Vol. 14. London: Macmillan and Free Press, 1968.

Lakatos, I. "History of Science and Its Rational Reconstruction." In R. C. Buck and R. S. Cohen (Eds.), *Boston Studies in the Philosophy of Science.* Vol. 8. Dordrecht, Netherlands: D. Reidel, 1971.

Lubrano, L. L. *Soviet Sociology of Science.* Columbus, Ohio: Anchor Press, 1976.

MacLeod, R. "Changing Perspectives in the Social History of Science." In I. Spiegel-Rösing and D. de S. Price (Eds.), *Science, Technology and Society: A Cross-Disciplinary Perspective.* Beverly Hills, Calif.: Sage, 1977.

Menard, H. W. *Science: Growth and Change.* Cambridge, Mass.: Harvard University Press, 1971.

Mendelsohn, E. "The Emergence of Science as a Profession in Nineteenth Century Europe." In K. Hill (Ed.), *The Management of Scientists.* Boston: Beacon Press, 1964.

Merton, R. K. *Science, Technology and Society in Seventeenth-Century England* [1938]. New York: Harper & Row, 1970.

Merton, R. K. *The Sociology of Science: Theoretical and Empirical Investigations.* Chicago: University of Chicago Press, 1973.

Merton, R. K. "The Sociology of Science: An Episodic Memoir." In R. K. Merton and J. Gaston (Eds.), *The Sociology of Science in Europe.* Carbondale: Southern Illinois University Press, 1977.

Merton, R. K., and Gaston, J. (Eds.). *The Sociology of Science in Europe.* Carbondale: Southern Illinois University Press, 1977.

Mulkay, M. J. "Some Aspects of Cultural Growth in the Natural Sciences." *Social Research,* 1969, *36,* 22–52.

Mulkay, M. J. *The Social Process of Innovation.* London: Macmillan, 1972.

Multhauf, R. P. "Reflections on Half a Century of the History of Science Society. II: The Society and Its Concerns." *Isis,* 1975, *66,* 454–467.

Oleson, A., and Brown, S. C. *The Pursuit of Knowledge in the Early American Republic.* Baltimore, Md.: Johns Hopkins University Press, 1976.

Ornstein, M. *The Role of Scientific Societies in the Seventeenth Century* [1913]. Chicago: University of Chicago Press, 1938.

Parker, J. "Comment on 'Three Models of Scientific Development' by M. J. Mulkay." *Sociological Review,* 1975, *23,* 527–533.

Price, D. J. de S. *Little Science, Big Science.* New York: Columbia University Press, 1963.

Price, D. de S. "Is Technology Historically Independent of Science?" *Technology and Culture,* 1965, *6,* 553–568.

Price, D. de S. "Citation Measures of Hard Science, Soft Science, Technology, and Nonscience." In C. E. Nelson and D. K. Pollock (Eds.), *Communication Among Scientists and Engineers.* Lexington, Mass: Heath, 1970.

Price, D. de S. "Comments on the Observations [On the Present Status of the History of Science]." *Isis,* 1975, *66,* 470–472.

Purver, M. *The Royal Society: Concept and Creation.* Cambridge, Mass.: M.I.T. Press, 1967.

Pyenson, L. "'Who the Guys Were': Prosopography in the History of Science." *History of Science,* 1977, *15,* 155–188.

Ravetz, J. R. *Scientific Knowledge and Its Social Problems.* New York: Oxford University Press, 1971.

Schofield, R. E. *The Lunar Society of Birmingham.* Oxford, England: Clarendon Press, 1963.

Shapin, S., and Thackray, A. "Prosopography as a Research Tool in the History of Science: The British Scientific Community, 1700–1900." *History of Science,* 1974, *12,* 31–74.

Sinclair, B. *Philadelphia's Philosopher Mechanics: A History of the Franklin Institute, 1824–1865.* Baltimore, Md.: Johns Hopkins University Press, 1974.

Spiegel-Rösing, I. S. *Wissenschaftsentwicklung und Wissenschaftssteuerung.* Frankfurt am Main: Athenäum, 1973.

Spiegel-Rösing, I. S. "The Study of Science, Technology and Society (SSTS): Recent Trends and Future Challenges." In Spiegel-Rösing, I. S. and Price, D. de S. (Eds.), *Science, Technology and Society: A Cross-Disciplinary Perspective.* Beverly Hills, Calif.: Sage, 1977.

Storer, N. W. *The Social System of Science.* New York: Holt, Rinehart and Winston, 1966.

Storer, N. W. "Introduction." In R. K. Merton, *The Sociology of Science.* Chicago: University of Chicago Press, 1973.

Thackray, A. "Natural Knowledge in Cultural Context: The Manchester Model." *American Historical Review,* 1974, *79,* 672–709.

Thackray, A. "Reflections on Half a Century of the History of Science Society. I: Five Phases of the History, Depicted from Diverse Documents." *Isis,* 1975, *66,* 445–453.

Thackray, A. "Measurement in the Historiography of Science." In Y. Elkana and others (Eds.), *Toward a Metric of Science.* New York: Wiley-Interscience, 1978.

Thackray, A., and Merton, R. K. "On Discipline Building: The Paradoxes of George Sarton." *Isis,* 1972, *63,* 473–495.

Webster, C. *The Great Insaturation: Science, Medicine and Reform, 1626–1660.* London: Duckworth, 1975.

Whitley, R. D. "Black Boxism and the Sociology of Science: A Discussion of the Major Developments in the Field." In P. Halmos (Ed.), *The Sociology of Science.* Sociological Review Monograph No. 18. Keele, England: University of Keele, 1972.

Woolgar, S. W. "Writing an Intellectual History of Scientific Development: The Use of Discovery Accounts." *Social Studies of Science,* 1976, *6,* 395–422.

Zuckerman, H. "Patterns of Name Ordering Among Authors of Scientific Papers: A Study of Social Symbolism and Its Ambiguity." *American Journal of Sociology,* 1969, *74,* 276–291.

Zuckerman, H. *Scientific Elite: Nobel Laureates in the United States.* New York: Free Press, 1977.

8

Derek de Solla Price

Ups and Downs in the Pulse of Science and Technology

My working hypothesis in this chapter (see also Price, 1978) is that the objectivity and transnational character of basic science lend to its historical development a much larger element of determinism and of imperviousness to local socioeconomic factors than one is accustomed to elsewhere in human affairs. It follows that a vital task of the historian of science and technology is to analyze such quasi-automatic secular change as it proceeds regardless of particular causes; only then can we dissect out those nonautomatic and significant events that require special ad hoc explanation. We need to perceive and understand regularity of behavior before we can get to second-order explanation of the deviations therefrom.

Taking the grand sweep in statistical historiography to measure the pulse, so to speak, of particular sciences and for particular countries is an old and doubtless influential tradition in our field. The importance of the early work by Rainoff (1929), Sorokin

Note: A version of this chapter was presented at the International Symposium on Quantitative Methods in the History of Science, Berkeley, Calif., Aug. 25–27, 1976.

(1937), and others has been well summarized by Merton (1977). Yuasa (1974) and Tomita and Hattori (1970, 1972, 1973, 1974) have added a wealth of data from Oriental sources, and Simonton (1975a, 1975b, 1975c, 1975d, 1975e, 1976a, 1976b) has given the data the more sophisticated statistical methodology that has long been needed for reliability estimates. All these previous studies have sought to squeeze from quantitative data the maximum information on the changing deployment of science and technology from field to field and from nation to nation. The difficulty is large and the intrinsic errors considerable, for one is trying to evaluate many different causes acting in concert, and these are seen against a background of random fluctuation that is huge in the crude data. Statistically, as a rule of thumb, if one generates data by counting a population of events sorted into pigeonholes, a count of N events implies a fluctuation of magnitude $\pm\sqrt{N}$. If one has as many as one hundred events in a group, the random error is therefore 10 percent; and one cannot be sure of the reality of distinctions from one group to another beyond such a limitation.

To avoid such problems, the present research demands only the minimum from the data, the main trend of an overview of all scientific and technological activity without any disaggregation by nation or by field. Thus, the maximum number of events in each annual group is ensured. To make the fluctuations even less, we have combined many sources of data and applied well-known smoothing procedures to the time series.

The starting point was an examination and hand count, page by page, through the following chronologies and histories of science and technology (extending from 1904 to 1966), recording each event that was given a precise or an approximate date:

- Ludwig Darmstädter, *Handbuch zur Geschichte der Naturwissenschaften und der Technik* [*Handbook for the History of Natural Science and Technology*]. (2nd ed.) Berlin, 1908.
- Ludwig Darmstädter and R. Du Bois-Reymond, *4000 Jahre Pionier-Arbeit in den Exakten Wissenschaften* [*Four Thousand Years of Pioneering Work in the Exact Sciences*]. Berlin, 1904.
- Franz M. Feldhaus, *Lexikon der Erfindungen und Entdeckungen auf den Gebieten der Naturwissenschaften und Technik in chronologischer*

Übersicht mit Personen und Sachregister [*Encyclopedia of Inventions and Discoveries in the Fields of Natural Science and Technology in Chronological Arrangement with Name and Subject Index*]. Heidelberg, 1904.

- Fielding H. Garrison, *An Introduction to the History of Medicine with Medical Chronology, Suggestions for Study and Bibliographic Data.* (4th ed.) Philadelphia and London, 1929.
- H. T. Pledge, *Science Since 1500: A Short History of Mathematics, Physics, Chemistry, Biology.* London, 1939.
- Paul Walden, *Chronologische Übersichtstabellen zur Geschichte der Chemie von den ältesten Zeiten bis zur Gegenwart* [*Chronological Tables for the History of Chemistry from the Oldest Times to the Present*]. Berlin, Göttingen, Heidelberg, 1952.
- Neville Williams, *Chronology of the Modern World: 1763 to the Present Time.* New York, 1966.

For control purposes a similar but smaller study was made of general history.

We then smoothed the year counts by taking a running weighted mean, the weighting factors being given by $\cos^2(90°\text{x}m/4)$, $m=0, \pm1, \pm2, \pm3$), so that it ran for three years on either side of the target year. Zeros being absent from the smoothed series, the logarithms of this mean were taken, a linear regression against time was computed, and the deviations from the regression were found. These deviations then gave for each source the amount by which it exceeded or fell short of a linear increase in the logarithm, that is, of regular exponential growth which is the gross deterministic behavior. The deviations from all sources were then averaged to give the results shown in Figure 2, and this was further given a grand smoothing by a running weighted average, this time taking in ten years on either side of target, to give the final product (shown as Figure 1). The data were found to be sufficiently numerous for some confidence from 1500 to the present time, though any individual source yielded information only over part of that range. A similar investigation for the general history control confirmed that the regularities of Figure 1 were peculiar to the history of science and technology; systematic fluctuations of this sort were far less well marked and quite different for the general history events.

Each particular history or chronology had, as expected, its

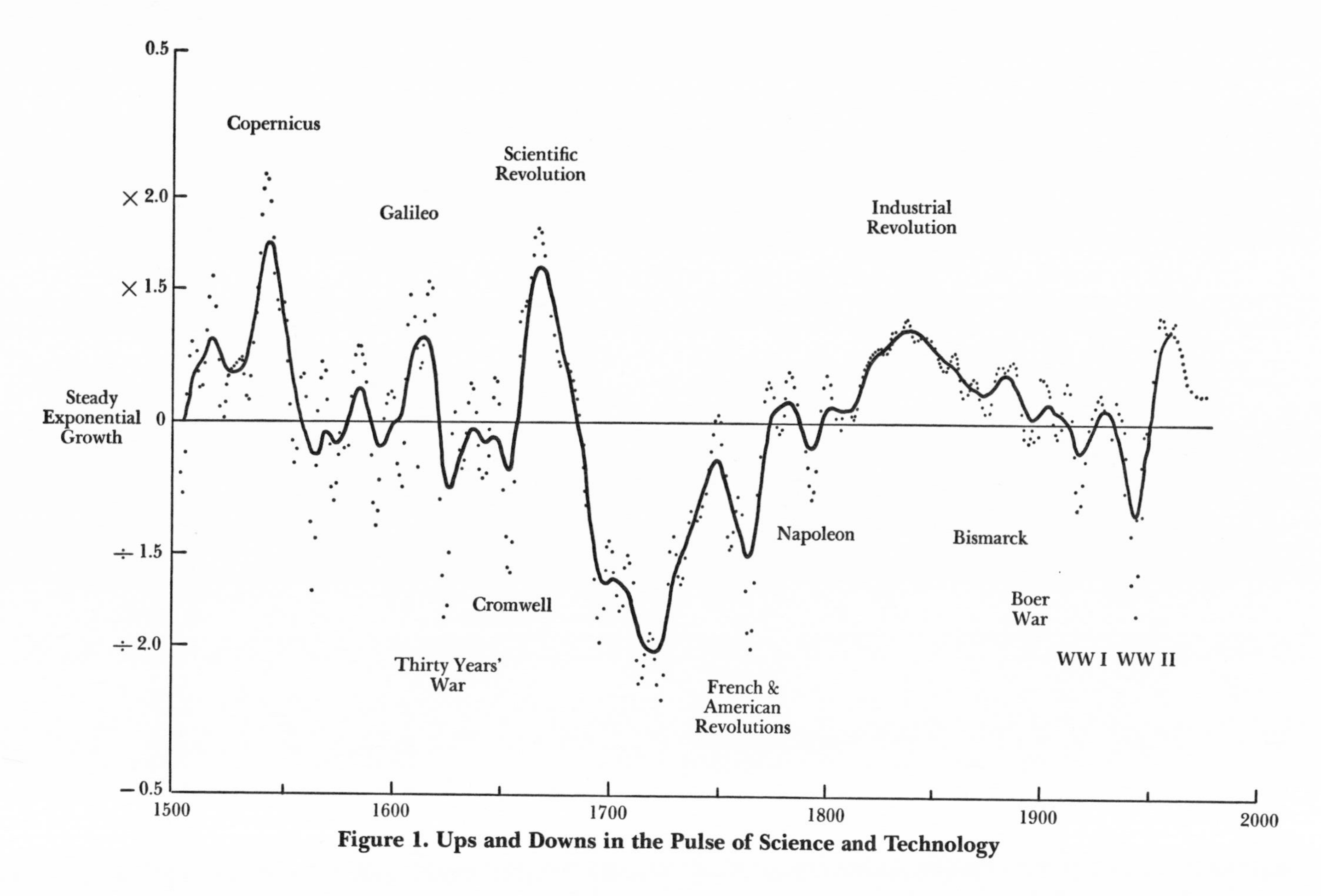

Figure 1. Ups and Downs in the Pulse of Science and Technology

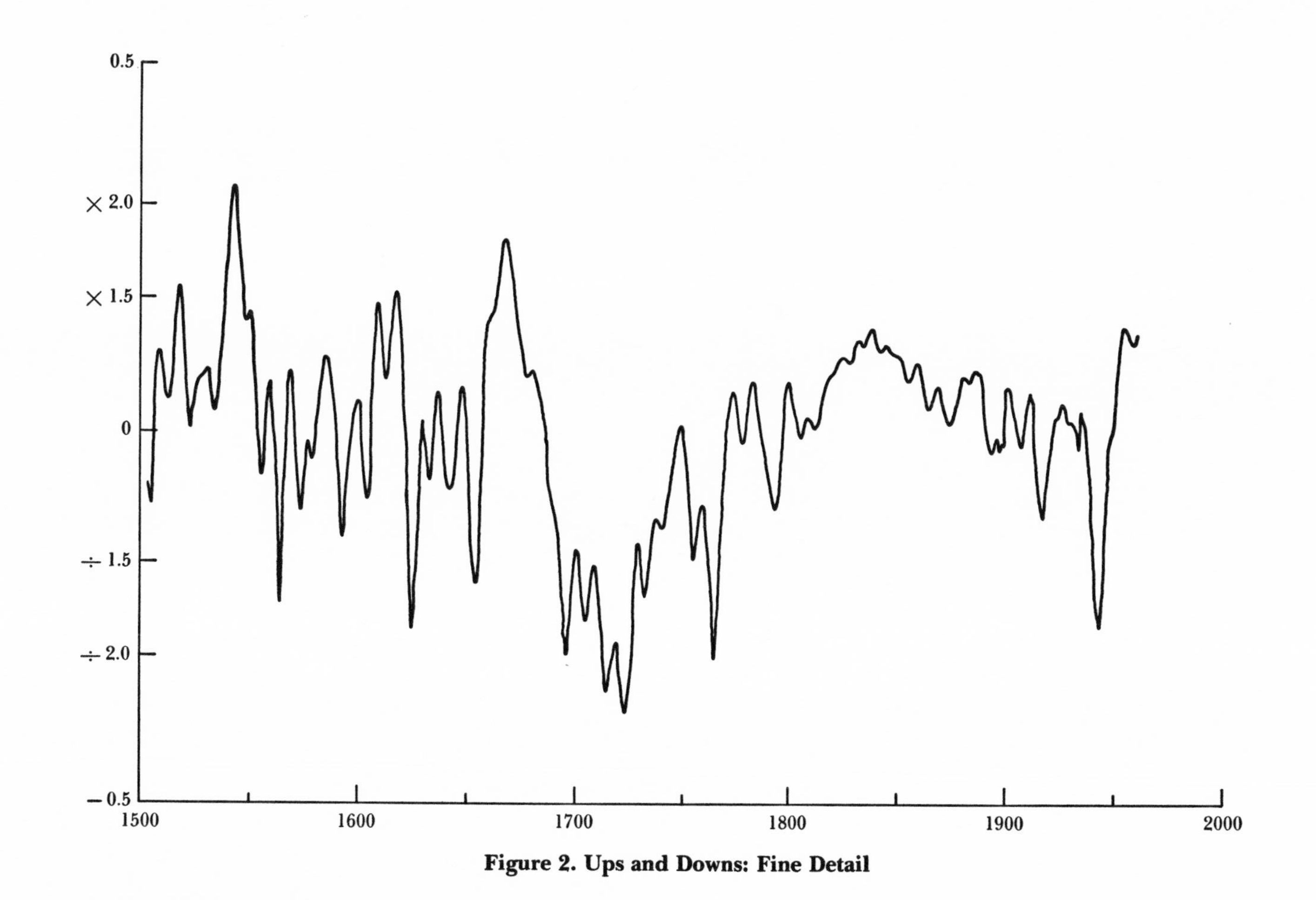

Figure 2. Ups and Downs: Fine Detail

own exponential growth rate, this depending partly on the acceleration of science itself and partly on the idiosyncratic perspective of the compiler or historian, concentrating more or less on the distant past compared with the more recent. The deviations of the logarithmic indicator, however, were very much the same from one source to another, correlating well and fitting together into an extensive time series that looks somewhat irregular when seen in the fine detail of Figure 2 but is much more systematic in the highly smoothed version of Figure 1.

It might be supposed that systematic deviation above and below steady exponential growth gives one the pulse of historians' interest rather than that of the events themselves. There exist, however, several statistical studies of particular scientific field bibliographies, and others of the total literature and the highly cited papers selected from it. These—for example, the careful study (see Figure 3) of comparative anatomy by Cole and Eales (1917)—show just the same pulse fluctuations; one is therefore driven to suppose that the present study indicates the intrinsic character of science rather than the prejudices and predilections of historians. For this reason we have been able to use recent citation data (Figure 4) to extend the historical study to the present day (dotted line on Figure 1).

The indicator thus derived charts the secular and systematic swings of the exponential growth *rate* of science; it is not a measure of the *quantity* of discovery and invention. It may be rather surprising that, over the entire range, the rate does not vary by more than about a factor of two above or below the average long-term trend which we expect on theoretical grounds. Within these limits, I maintain that the indicators give one a measure of scientific activity that agrees well with the historians' intuition of relatively active and inactive periods.

Among the obvious expectations for eras of high activity, one finds indicated very clearly the peaks of the scientific revolution and the industrial revolution as well as localized outbursts in the times of Copernicus and Galileo (I use the names only as surrogates for their periods). The major wars and social upheavals of history correspond to localized troughs that are very clear.

Highly unexpected (at least to one historian) is the enormous trough indicating a post–scientific revolution slump of more monumental proportions than any other event, and the rather

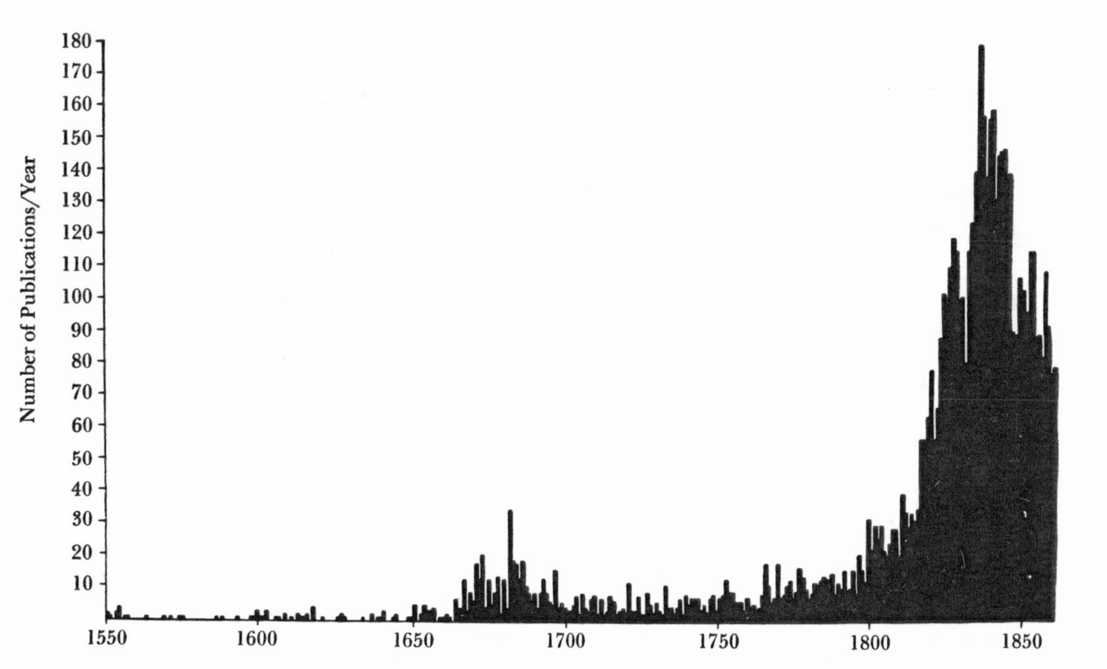

Figure 3. Comparative Anatomy Publication Between 1540 and 1860 (adapted from Cole and Eales, 1917).

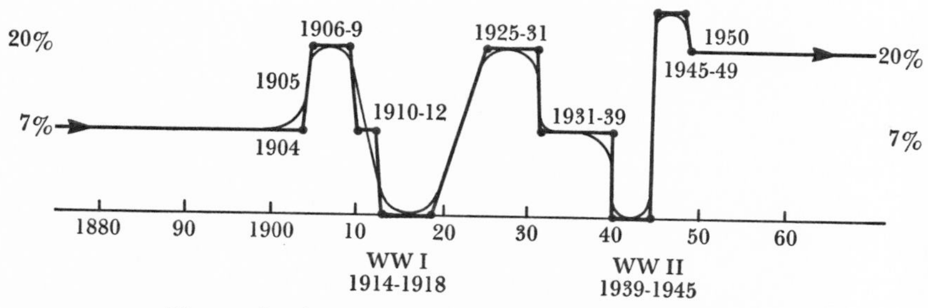

Figure 4a. Acceleration of Scientific Production.

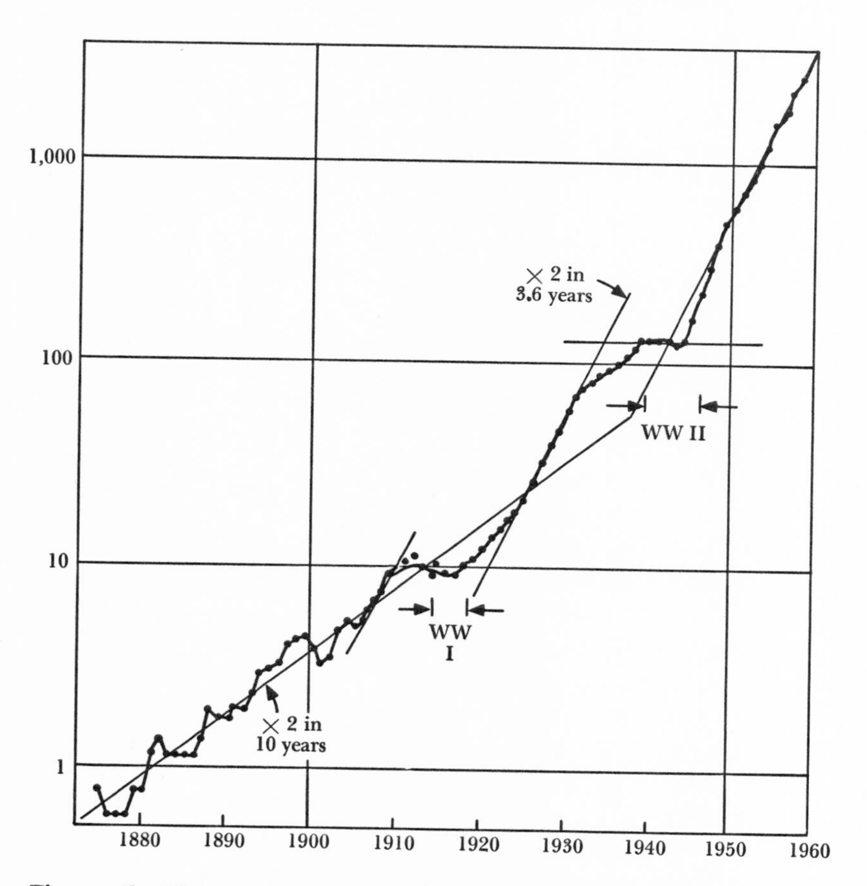

Figure 4b. Five-year Running Average of Number of Heavily Cited Papers per Year by Year of Publication. All Papers Cited at Least Ten Times in Any Single Year During the Period 1961–1972.

weak and diffuse low profile character of the Industrial Revolution compared with other peaks. From the first, one must suppose that more is at stake than Newton being a hard act to follow; from the second, it seems that the industrial revolution may be a compartmentalizing convenience of the historian rather than an actuality.

I take the conventional view of historians of science to be that, ancient and medieval groundings notwithstanding, modern science began with bursts of activity in the fifteenth and sixteenth centuries, which led to the seventeenth-century scientific revolution, setting the tone for all later developments, including a postponed revolution in chemistry and still other postponed events in other fields. These new quantitative data for the general secular changes make me revise this convention. That is, I now assume that the early isolated peaks for Copernicus, Galileo, *and* the scientific revolution represent precursor-like events and that the main trend leading to modern times began only toward the end of the eighteenth century. I suggest therefore that the (Lavoisier) revolution in chemistry is not a "postponed" event but a marker of a period when the modern movement would have had its "natural" and expected beginning. There was therefore a premature inception through the accident that astronomy comes out surprisingly neatly and had already been successfully mastered by Ptolemy and the followers of mathematical astronomy. In other words, chemistry and biology were *not late;* astronomy and mechanics were *early* and then suffered a lag during the comparatively dull first half of the eighteenth century.

The most recent period is best understood, in terms of this general indicator, as a somewhat typical postwar recovery from a trough, leading to an overshooting and a contemporary recovery to oscillation in the neighborhood of the long-term average growth rate. Thus, the modulation of steady exponential growth seems to correspond reasonably well with both historical and modern intuitive estimates of general scientific and technological activity.

References

Cole, F. J., and Eales, N. B. "The History of Comparative Anatomy, Part I." *Science Progress,* 1917, *11* (44) 578–596.

Merton, R. K. "The Sociology of Science: An Episodic Memoir." In R. K. Merton and J. Gaston (Eds), *The Sociology of Science in*

Europe. Carbondale: Southern Illinois University Press, 1977.
Price, D. de S. "Toward a Model for Science Indicators." In Y. Elkana and others (Eds.), *Toward a Metric of Science: The Advent of Science Indicators.* New York: Wiley-Interscience, 1978.
Rainoff, T. J. "Wave-Like Fluctuations of Creative Productivity in the Development of West-European Physics in the Eighteenth and Nineteenth Centuries." *Isis,* 1929, *12,* 287–319.
Simonton, D. K. "Invention and Discovery Among the Sciences: A P-Technique Factor Analysis." *Journal of Vocational Behavior,* 1975a, *7,* 275–281.
Simonton, D. K. Private communication, Sept. 12, 1975b.
Simonton, D. K. Private communication, Sept. 19, 1975c.
Simonton, D. K. Private communication, Oct. 5, 1975d.
Simonton, D. K. "Sociocultural Context of Individual Creativity: A Transhistorical Time-Series Analysis." *Journal of Personality and Social Psychology,* 1975e, *32,* 1119–1133.
Simonton, D. K. "The Causal Relation Between War and Scientific Discovery: An Exploratory Cross-National Analysis." *Journal of Cross-Cultural Psychology,* 1976a, *7,* 133–144.
Simonton, D. K. Private communication, Jan. 3, 1976b.
Sorokin, P. A. *Social and Cultural Dynamics.* New York: American Book Co. 1937.
Tomita, T., and Hattori, K. "History of Science Society of Japan (Ed.): *Nihon Kagaku-Gijutsu-Shi Taikei* [*History of Science and Technology in Japan*], 25 vols., 1964–1970." Book review. *Japanese Studies in the History of Science,* 1970, *9,* 164–167.
Tomita, T., and Hattori, K. "Compilation of a Thesaurus and Total Index for *Nihon Kagaku-Gijutsu-Shi Taikei* by Means of a Computer." *Japanese Studies in the History of Science,* 1972, *11,* 41–65.
Tomita, T., and Hattori, K. "Outline of a Thesaurus of *Nihon Kagaku-Gijutsu-Shi Taikei* with Heading List of Classified Key Words." *Japanese Studies in the History of Science,* 1973, *12,* 15–38.
Tomita, T., and Hattori, K. "Some Considerations of Quasi-Quantitative Analysis of the History of Science in Japan by Key Words—Trial of Quantitative History." Paper presented at International Congress of the History of Science, Tokyo, 1974.
Yuasa, M. "The Shifting Center of Scientific Activity in the West." In S. Nakayama and others (Eds.), *Science and Society in Modern Japan.* Tokyo: University of Tokyo Press, 1974.

9

Nico Stehr

The Ethos of Science Revisited

Social and Cognitive Norms

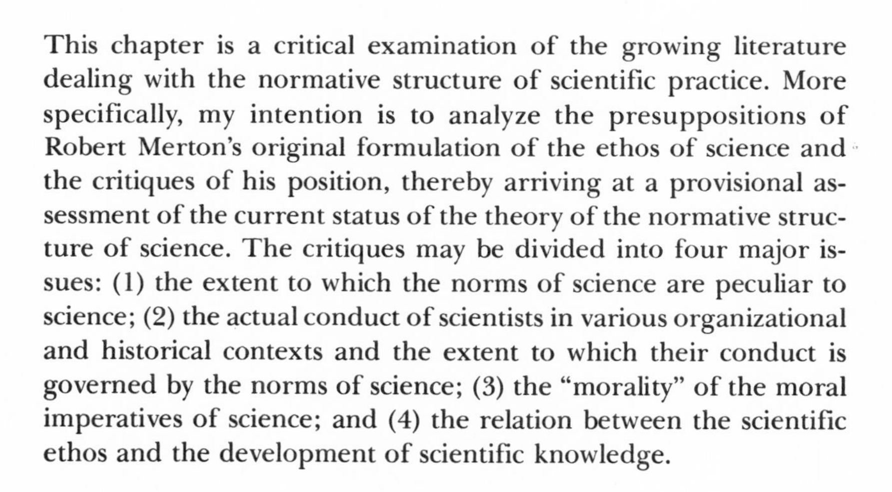

This chapter is a critical examination of the growing literature dealing with the normative structure of scientific practice. More specifically, my intention is to analyze the presuppositions of Robert Merton's original formulation of the ethos of science and the critiques of his position, thereby arriving at a provisional assessment of the current status of the theory of the normative structure of science. The critiques may be divided into four major issues: (1) the extent to which the norms of science are peculiar to science; (2) the actual conduct of scientists in various organizational and historical contexts and the extent to which their conduct is governed by the norms of science; (3) the "morality" of the moral imperatives of science; and (4) the relation between the scientific ethos and the development of scientific knowledge.

Note: I would like to acknowledge a critical reading of the first draft of the essay by Michael Kozlowski and the helpful editorial assistance of Laura Hargrave.

The Normative Structure of Science

Critics and supporters of Merton's theoretical formulation of the moral imperatives of science seem to agree that the formulation provided one of the most significant theoretical foundations for a sociology of science (see Barnes and Dolby, 1970, p. 3; Spiegel-Rösing, 1973, p. 37; Storer, 1973, p. 226; Mitroff, 1974a, pp. 12–13). A "first" delineation of the normative structure of science by Merton appears in the context of his essay "Science and the Social Order," initially presented as a paper to the American Sociological Society in 1937 and published in 1938 in *Philosophy of Science.* In the context of this essay, Merton ([1938] 1973d, p. 259) observes, "The sentiments embodied in the ethos of science [are] characterized by such terms as intellectual honesty, integrity, organized skepticism, disinterestedness, impersonality." In the same essay, one also finds references to psychological attributes that scientists share with nonscientists—attributes that also affect their scientific works. Merton's first systematic exposition, however, is found in a 1942 essay, originally published under the title "A Note on Science and Democracy" in the *Journal of Legal Political Sociology.* Subsequently, the same essay appeared as "Science and Democratic Social Structure" in Merton's *Social Structure and Social Theory* (1949). Still later, the essay was reprinted as "The Normative Structure of Science" in Merton's *The Sociology of Science: Theoretical and Empirical Investigations* (1973). In the course of these several reprintings, the essay has not remained unchanged. Most of the changes, however, involve minor matters, such as stylistic improvements.

The ethos of science generally is conceptualized by Merton ([1942] 1973b, pp. 268–269) as that "affectively toned complex of values and norms which is *held to be binding* on the man of science. The norms are expressed in the form of prescriptions, proscriptions, preferences, and permissions. They are legitimized in terms of institutional values. These imperatives, transmitted by precept and example and *reinforced by sanctions* are in *varying degrees internalized by the scientist.* . . . Although the ethos of science has not been codified, it can be *inferred* from the moral consensus of scientists as *expressed in use and wont,* in countless writings on the scientific spirit and *in moral indignation directed toward contraventions of the ethos*" (italics added).

Merton specifies four basic moral imperatives, which provide a foundation for the social relations of science, and the professional identity of individual scientists and therefore constitute important elements of the sociocultural structure of science.

1. *Universalism* prescribes that knowledge claims in science should be evaluated and accepted or rejected according to impersonal cognitive criteria rather than the "personal or social attributes of their protagonist" (Merton, [1942] 1973b, p. 270) and that careers and opportunities in science should be based on achievement and competence only. (The choice of words perhaps is significant; the expression "should be" denotes a prescription aspired to but not necessarily a description of consistent scientific practice.)

2. *Communism*[1] refers to the interrelated "communal" (public) character of scientific knowledge claims; the corresponding limited "rights" of the originator(s) to recognition and esteem, resulting in the distinctive and anomalous character of intellectual *property* in science;[2] and the imperative not to withhold knowledge claims—an imperative reinforced by the "incentive of recognition, which is, of course, contingent upon publication" (Merton, [1942] 1973b, p. 274).

3. *Disinterestedness*, the moral imperative at the institutional level of science, is largely self-explanatory (see Parsons, 1939; Wunderlich, 1974); it points to a distinctive structure of control exercised over the individual motives of scientists.

4. *Organized Skepticism* is "both a methodological and an institutional mandate" (Merton, [1942] 1973b, p. 277). Knowledge claims should not be accepted without (socially organized) scrutiny but should be warranted with reference to the technical norms of science.

In addition, Merton ([1942] 1973b, pp. 269–270) refers to certain technical (or methodological) norms, certified knowledge (the institutional goal of science), and institutional values as other elements of the culture of science. Certain other norms—for instance, "individualism" and "rationality" (Barber, 1952, pp. 86–90; Barnes and Dolby, 1970, p. 9) "objectivity" and "generalization"

(Storer, 1966, p. 81), and various other formulations (see, for example, Cournand and Zuckerman, 1970; Cournand, 1977)—have been suggested as a part of the ethos of science; these norms were intended to be largely compatible with the original formulation advanced by Merton.

Another norm mentioned by Merton [1957] 1973c, p. 294) is the "emphasis upon originality on the institutional plane" in the scientific community; as emphasized in the context of the same essay, originality counterbalances humility, thereby creating potential normative conflict ([1957] 1973c, pp. 303, 305; see also Merton, 1976). It has often been observed, therefore, that science as a social institution is, like any other societal institution, characterized by potentially incompatible normative demands (norms and counternorms). (A discussion of the sociostructural antecedents and consequences of norms and counternorms in social organizations may be found in Merton, 1976.) As long as potentially conflicting norms are not "compartmentalized" (Deutscher, 1972), they may generate "ambivalence"; for example, ambivalence toward claims of priority, particularly in the context of multiple independent discoveries. Finally, Merton emphasizes that there have been comparatively few empirical instances of deviant responses to particular norms of science (Merton, [1957] 1973c, p. 321), and he stresses his theory of social structure and anomie applies also to the institution of science (Merton, [1957] 1973c, p. 308, note 51).

The four basic moral imperatives of science are more than mere moral principles. They are linked in distinct ways to the cognitive development of science. For, as Merton ([1942] 1973b, p. 270) notes, "The mores of science possess a methodologic rationale but they are binding, not only because they are procedurally efficient, but because they are believed right and good. They are moral as well as technical prescriptions." (Luhmann, [1968] 1970, writing from a functional-structural perspective, has described the functional necessity for *social* mechanisms to operate as *cognitive* mechanisms in science.) Thus, a definite correlation between moral imperatives and the advancement of scientific knowledge is implied. For instance, "objectivity precludes particularism. The circumstance that scientifically verified formulations refer in that specific sense to objective sequences and correlations militates against all efforts to impose particularistic criteria of validity" (Merton, [1942] 1973b, p. 270.) In other words, the norms do not merely regulate the behavior or the social relations of members of the scientific community; also, in distinct ways, they enhance the institutional goal of science, which is the continuing extension of certified knowledge claims.[3]

It is therefore clear that, in Merton's view, the norms of science derive from the purpose of science: "The institutional imperatives (mores) derive from the goal and the methods. The entire structure of technical [cognitive] and moral norms implements the final objective" (Merton, [1942] 1973b, p. 270). The normatively prescribed social relations of science therefore complement, if not implement, the development of knowledge in science and vice versa. The dialectics of social and cognitive norms are a part of the institution of science (for a different view, see Bourdieu, 1975; Knorr, 1977). However, an implicit bifurcation of social and cognitive norms is equally evident when Merton outlines the dialectics of social and cognitive norms as constitutive of science. His emphasis then shifts to a more detailed inquiry into the normative basis of the social relations of science and therefore to a dissociation of the interdependent realms of intellectual and social domains of science. The former emphasis may be called the *theoretical* or *programmatic* formulation of the ethos of science; the latter, more restrictive, formulation may be called the *research* formulation of the ethos of science (followed subsequently by many sociologists of science; see Cole and Zuckerman, 1975, p. 157). The two formulations—differing in scope as they do—seem, on the one hand, to represent a significant *theoretical break* in the work of Merton and therefore in the sociology of science. But, on the other hand, they are often seen as a theoretical basis and justification for empirical research into the social relations of science. The observed theoretical break cannot be seen, in a historical context, to represent the sole condition for such categorical bifurcation; the then dominant philosophies and histories of science, as well as the dominant societal legitimations of scientific practice also support it[4] (King, 1971).

In addition to the thesis that social relations of science based on these norms support the advancement of scientific knowledge, the 1942 essay also refers to potential interinstitutional normative support and conflict.[5] For example, a societal legitimation of ascriptive standards for careers would contradict the norm of universalism in the ethos of science.

Merton's normative structure of science and the subsequently developed theory of the reward system in science represent the structural-functional account of the social relations of science. In the course of their professional socialization into science (Merton, [1942] 1973b, p. 269), scientists acquire the ethos of science; it then becomes a salient part of their identity and consciousness and is reinforced through mechanisms of social control, mediated by the reward system of science. Thus, "rewards are to be meted out in accord with the measure of accomplishment. When the institution of science operates effectively, the augmenting of knowledge and the augmenting of personal fame go hand in hand" (Merton, [1957] 1973c, p. 323).

Merton's formulation obviously contains a number of presuppositions, many of them noted later by critics. Merton assumed a particular historiography of science and a particular theory of the institutional setting of scientific practice. The ethos of science is assimilated, on the one hand, to a particular theory and history of knowledge and, on the other hand, to a particular model of the social relations of science—a model that emphasizes the normative basis and control of social action. In each instance, the relative compatibility of cognitive and social elements within science is emphasized. Threats to such compatibility primarily come from sources external to science such as the state. Cognitive divisions therefore are deemphasized,[6] because the theory of the institutional basis of science and its cognitive development reinforce each other. Yet, one problem that causes considerable discussion is the precise methodological status of Merton's formulation: Is it to be understood as an "ideal-type" or is it perhaps a "general conception" that claims to have captured the actual, observable conduct of scientists?[7] But the most significant controversial legacy of the Mertonian conception of the ethos of science is linked to the theoretical break. Critics of the ethos of science have reproduced the break, although often on the basis of different intellectual traditions and presuppositions.

Critiques of the Ethos of Science

What, then, have been the serious objections to Merton's formulation of the ethos of science? Most of the criticisms can be placed in one of several categories.

First, and perhaps most often raised because it directly pertains to both a particular historiography of science and a theory of the social relations of science, is the question: Are the norms of science peculiar to science? A related question is the extent to which the norms actually affect the conduct of scientists. A third question is whether the moral imperatives of scientific practice are indeed moral. Finally, questions have been raised about the claim of a correlation between the ethos of science and the development of scientific knowledge.

Most of the critiques in each of the categories can be classified as post-Kuhnian critiques; that is, they reflect the immense influence generated by Kuhn's *The Structure of Scientific Revolutions* (1962). Kuhn provides critics with an alternative historiography of scientific knowledge and, tacitly, an alternative theory of the social relations of science. Many of the criticisms to be discussed were made possible by a particular reconstruction of Kuhn's account of scientific practice (Barnes and Dolby, 1970, p. 3; Spiegel-Rösing, 1973). Some examples to illustrate the presuppositions are appropriate: "The paradigm is a source of *social control*" (Barnes and Dolby, 1970, p. 197; emphasis added). "The cohesion, solidarity and commitment within [paradigm-sharing communities] stem from the technical norms of paradigms, not from an overall scientific 'ethos'" (Barnes and Dolby, 1970, p. 23; see also Mulkay, 1972b, p. 15). These researchers deal with Merton's formulation primarily with reference to its cognitive functions (Barnes and Dolby, 1970, p. 4; King, 1971, pp. 15–16; Mitroff, 1974a, pp. 10, 13). In contrast, researchers who do not make any explicit reference to Kuhn's historiography of scientific knowledge typically describe the ethos of science as a set of *social* norms. For instance, Hagstrom (1965, p. 1) writes, "I am concerned with the operation of social control within the scientific community, with the problem of discovering the social influences that produce conformity to scientific norms and values" (see also West, 1960, p. 54; Rothman, 1972).

The resulting shift of domain assumptions, however, does not produce a negation of the theoretical break observed earlier. The emphasis of research shifts to an analysis of cognitive processes in science and therefore represents a separation of social and intellectual factors in scientific practice.

The general structure of these critiques is as follows. Inasmuch as (the predominant reconstruction of) Kuhn's historiography of the development of scientific knowledge provides an important foundation for such critiques, the criticism shifts to the *cognitive* development of science, including such questions as: To what extent is the ethos of science governed by the cognitive development of science? Or, even stronger: To what degree are the actual norms of science cognitive (technical) norms? Not infrequently, therefore, the cognitive norms are conceptualized, in some sense, as prior to the internalization of the social norms of science. The acquisition of social norms is seen as mediated by the conceptual structure of science (Kuhn, 1962, p. 46; Downey, 1967, p. 251). Obviously, such considerations have implications for a theory of the social relations of science. In contrast, those who choose not to emphasize the theoretical consequences of Kuhn's historiography of scientific knowledge for the ethos of science typically inquire into the *social* relations of science in light of the normative structure of science and are therefore not bound to an analysis of the "rational" structure and content of scientific discourse (see, for example, Gouldner, 1976). The overall result has been a perpetuation of the bifurcation of theory and inquiry into the social and cognitive norms of science. Few analyses have systematically investigated the questions raised by the observation that "methodological canons are often *both* technical expedients and moral compulsives" (Merton, [1942] 1973b, p. 268; emphasis added).

1. *Are the norms of science peculiar to science?* How are we to differentiate science as a social and cultural institution from other social and cultural institutions? Any criterion or set of criteria found acceptable depends on an implicit or explicit theory of social and cultural differentiation. One could, for example, emphasize the differentiation of modes of production, relations of production, technology, social functions, norms and values, types of knowledge, modes of discourse, or a combination of these. By vir-

tue of its focus on "a set of cultural values and mores governing the activities termed scientific," Merton's ([1942] 1973b, p. 268) formulation of the ethos of science links his implicit theory of social differentiation to considerations that emphasize the distinctiveness (but not necessarily radical discontinuity) of norms in different institutions.

Most of Merton's critics do not deny that norms, in some sense, are fundamental to the differentiation of social institutions; but many of them believe that Merton's formulation of the normative structure of science does not adequately set science, as an institution, apart from other social institutions. Barnes and Dolby (1970, pp. 8–12), for example, observe that the norms of universalism, rationality, and organized skepticism are not peculiar to science but are also observable in everyday life. Similarly, Mulkay (1969, pp. 22, 27) questions the adequacy of the ethos of science as a valid means of differentiating science. He suggests instead that differentiation must be linked to a theory of the development of specialized knowledge claims: "The theoretical and methodological norms are more central to the structure of the scientific community than are the Mertonian social norms" (Mulkay 1969, p. 36; see also Mulkay, 1972a). Evidently, the dichotomy of social and cognitive norms is crucial to the argument, and the claim is advanced that the shared type of knowledge as well as the procedures (the technical-methodological rules) used to generate and warrant knowledge claims are constitutive of science. Accordingly, the identity of the scientific community is a cognitive identity (see also Parsons, 1951, pp. 336–337; Boehme, 1974). Mulkay's approach therefore is based on a theory of social differentiation that stresses the emergence of specialized cognitive procedures and their consequences.[8]

Thus, following Kuhn's historiography of scientific knowledge and his tacit theory of the social relations of science mediated by cognitive processes, the criticisms for the most part amount to an "alternative" delineation of a set of norms as constitutive of scientific practice. Kuhn's approach therefore is perceived as a reversal of Merton's approach (King, 1971, pp. 18, 30). In the introduction to his most recent publication, *The Essential Tension,* however, Kuhn (1977) appears to suggest "Je ne suis pas Kuhniste," when, for example, he insists that "scientific communities must be discovered by examining patterns of education and communication before asking which particular research problems engage each group" (p. xvi) and when he more generally stresses the continuity between his concerns and those of Merton (p. xxi). Nonetheless, most of the critiques of Merton's formulation remain based on a particular reconstruction of Kuhn's theory of scientific practice.

Because these critiques view scientific practice primarily as "conceptual," "cognitivist," or "discursive," they cannot cope successfully with the "material" conditions of scientific practice, particularly as such conditions affect the cognitive development of scientific knowledge. As a result, relative to the theoretical formulation of the ethos of science, they seem to represent a retrogressive step. With respect to the research formulation of the ethos of science, however, the critiques may be seen as progressive, since they emphasize the intellectual basis of scientific practice. Yet, the criticisms lack a sense of historicity (Boehme, van den Daele, and Krohn, 1972), but historical changes are difficult to detect once the issue is confined to cognitive norms.[9] Thus, the fundamental issue of the historicity and interdependence of social and cognitive domain in science (Fleck, 1935; Bourdieu, 1975, p. 22; Weingart, 1976, pp. 33–92; Baldamus, 1977; Knorr, 1977) remains a challenge to a comprehensive theory of the normative structure of science (Barnes and Dolby, 1970, pp. 14–18).

In this context other questions about the ethos of science derive from differing theories of social institutions (see Mitroff, 1974a, p. 77), particularly questions about the exact character of norms in institutions. As noted earlier, Merton has repeatedly stressed the presence, structural origin, and import of "contradictory" norms in social institutions, including science.[10] Mitroff (1974a, p. 17) also emphasizes the importance of such conflicting sets of norms as conditions for the possibility of rationality and growth in science.[11] Mulkay (1976, p. 641), however, has advanced the thesis that norms and counternorms, as discussed by Mitroff or Merton or a combination of the two, should not be seen as rules governing scientific practice—in particular, the construction of specialized knowledge claims. (For a critique of Mulkay's views, see Zuckerman, 1977, pp. 126–128; Gaston, 1978, pp. 178–180.)

Rather, Mulkay argues, specific intellectual commitments are a part of scientific practice. These commitments mediate sociostructural processes such as inequality and communications networks (Mulkay, 1976, p. 643; see also Kuhn, 1977, pp. ix–xxiii). Norms and counternorms, Mulkay suggests, are not institutionalized norms; they are not related, for example, to the exchange of rewards in science (Mulkay, 1976, pp. 641–645): "Conformity to most of the supposed norms and counternorms of science is largely irrelevant to the institutional processes whereby professional rewards are distributed" (Mulkay, 1976, p. 642). Instead, and as Mulkay has argued elsewhere, operative control mechanisms are intellectual commitments.[12] According to Mulkay, scientists may refer to norms and counternorms when describing their practice—for instance, in conversations—but that does not prove the existence of a functioning normative structure in science, which, for example, may be said to contribute to the advance of scientific knowledge.

2. *Are the norms prescriptive?* The question of the extent to which the various norms (and counternorms) actually determine or influence social and cognitive processes in science has been addressed in two different ways. Some critics relativize or limit the importance of the ethos of science. Others seek to determine the extent to which the norms of science operate empirically in different scientific communities.

Among those who attempt to limit the importance of the ethos of science, Mitroff (1974b, pp 585–586, 594) argues that its applicability may vary with the cognitive institutionalization of a research area, specialty, or discipline. A lack of cognitive institutionalization, (which may, of course, be a more or less permanent attribute of a research area or specialty) reduces the functional importance of the norms. Or, as Storer (1973, p. xiv), assimilating the historiography of scientific knowledge as developed by Kuhn, suggests, "It is when . . . a universe of discourse is only slightly developed (as during the Kuhnian 'preparadigm' stage in the development of a new discipline or during a 'scientific revolution'), or when group loyalties outside the domain of science take over, that violations of the norms become more frequent, leading some to reject the norms entirely." A similar conclusion may be found in Ben-David (1977) and Zuckerman (1977), who both indicate that the norms operate relative to the progress made in resolving a particular research problem; that is, the ethos of science is seen as predominantly the normative structure of the context of justification, not the context of discovery.

Similarly, critics of Merton's formulation have asked whether one should distinguish between two not necessarily congruent sets of norms; namely, "professed" and "statistical" norms. The latter type of norm actually regulates conduct among scientists, while the former is ritualistically "enjoined or celebrated in tract or speech" and "may be directed mainly to out-groups in situations of justification and conflict" (Barnes and Dolby, 1970, p. 8; see also Merton, 1976, pp. 40–41). In short, professed norms are described as an occupational ideology (Mulkay, 1976, p. 646). The scientist is someone in need of a dual morality: an idealized morality for the "external," legitimating, social relations and a realistic morality for the internal social relations of science (see Spinner, 1977, p 565). Moreover, the idealized norms may have originated in everyday life as characteristics of popular science (Fleck, 1935, p. 123; Spinner, 1977, p. 563). In the context of these objections, therefore, an important question concerns the extent of the congruence between the two sets of norms.

The validity of the ethos of science is also relativized by those who emphasize another social context: the social relations of applied versus pure science. These critics claim that the ethos of science is primarily the ethos of pure science (Barber, 1952, p. 95; Marcson, 1960; Krohn, 1961; Kornhauser, 1963; Storer, 1966, pp. 165–166; Box and Cotgrove, 1966; Barnes and Dolby, 1970, p. 7; Kowalewski, 1974, p. 279). Other observations designed to limit the ethos of science, or particular aspects of it, make reference to social constraints. Sklair (1973, p. 113), for example, observes that "much, if not most, contemporary science is carried out under conditions of formal or informal secrecy"—necessitated by national security matters and matters of economic interest (see also Buehl, 1974, p. 117). Finally, some arguments are directed toward a historical relativization of the ethos of science. These critics generally allege that the ethos of science may have been operative in the past, particularly in very early phases of the development of scientific institutions (Ravetz, 1971, p. 310), but is no longer operative.

However, those who attempt to relativize the ethos of science agree, in the last analysis, that scientific practice is indeed based on certain rules and standards after all. As Barnes and Dolby (1970, p. 10) therefore stress, "The scientist has a specific viewpoint which makes him skeptical of some results whilst uncritically accepting others; this is the norm." The social and cognitive relations of science, that is, do follow certain standards. At issue therefore are different formulations regarding exactly how and when the norms apply. A growing number of empirical investigations into the normative structure of science, for the most part influenced by the research formulation of the ethos of science, have dealt, at least in part, with this question. Gaston's (1978, p. 186) conclusion is perhaps representative of many of these empirical investigations: "The normative imperatives, universalism, communality, disinterestedness, and organized skepticism, are strongly institutionalized in the British and American scientific communities studied" (see also Cole and Cole, 1973).

3. *Are the moral imperatives moral?* Much less frequently discussed is the question of the morality of the ethos of science. Merton's original formulation did not address this question explicitly. In general, however, critics seem to have heeded Durkheim's warning that a discussion of the morality of morals will indicate only the morals of the moralist, not those practiced by others. Nevertheless, some critics refer to this issue in passing. Mok and Westerdiep (1974, p. 216) observe that scientists may use, perhaps abuse, the norm of organized skepticism and disinterestedness in order to stay aloof from public controversies. Similarly, Sklair (1973, pp. 111–112) claims that the norm of universalism, taken to the extreme, would ensure that "no squeamish, unscientific rules prohibiting crucial experiments on human subjects" would obstruct medical science. Both objections make evident that science cannot insulate its moral relations from society's moral views. Although the norms of science need not be the norms of society, society's norms affect those of science.

4. *Are the functional imperatives functional?* A final major issue about the norms of science concerns their influence on the growth of scientific knowledge. Practically no empirical research has been directly related to this question.[13] Any assessment of the degree to which the ethos of science claims to facilitate the production of scientific knowledge presupposes an explication of a historiography of scientific knowledge, and a specification of criteria for the growth of knowledge. Claims made about the effect of the social relations of science on the cognitive development of science obviously require some notion of the patterned development of knowledge. Moreover, the "functional" effects of the norms on the cognitive development of science must be either direct or mediated. In either case it is necessary to have a theory of the relation between social and cognitive processes within science. The ethos of science may have an indirect effect, if it provides one of the main conditions for organizing scientific discourse in a manner which allows for the advancement of knowledge. The effect may be more direct, for example, if the ethos of science is formulated in such a way that it already incorporates cognitive conditions for the possibility of growth of scientific knowledge claims.

Mitroff (1974a, 1974b) has explicitly questioned the validity of the notion that the ethos of science is functional for the advancement of scientific knowledge. Specifically, Mitroff (1974a, pp. 75–78) suggests that certain counternorms, such as the degree of secrecy or emotional commitment of scientists, may indeed further the progress of scientific knowledge.[14] But, in spite of Mitroff's view, a comprehensive examination of the relation between the normative structure of science and the development of scientific knowledge requires the explication of a comprehensive theory of the normative (social and cognitive) structure of science, a task yet to be accomplished.

Conclusion

One of the salient characteristics of the debate about the ethos of science is the categorical bifurcation of social norms and cognitive norms and the restrictive formulation of a theory of the normative structure of science. Merton's theoretical formulation of the ethos of science, in contrast to the research formulation and the cognitivist formulation of the critiques of the research conception of the ethos of science, is best suited to transcend "the difficulty of uniting into a single vision the perspectives of science as conceptual structure and as human activity" (Gutting, 1973, p. 209). The dichotomy between social and cognitive norms of science, and therefore the dichotomy between scientific ideas and scientific practice, clearly requires a return to the original theoretical formulation of the ethos of science.

Notes

1. I intend to retain the formulation *communism* since the *Zeitgeist* has undergone change once again and does not at the present time, so it seems, require a substitute term such as *communality* (see Barber, 1952, p. 91).

2. A rather similar observation is made by Ravetz (1971, p. 299): "One essential feature of the scientist's intellectual property, as distinguished from the 'real' property of commerce, is that it exists only by being available for use by others." The anomalous and distinctive character of intellectual property in science, as an essential context and expression of the norm of communism, also is emphasized in several of Merton's subsequent writings; more recently, for example, Merton (1977, pp. 48–49) notes that the "distinctively anomalous character of intellectual property, becoming fully established in the domain of science only by being openly given away (published), is linked with the normative requirement for scientists making use of that property to acknowledge (publish) the source, past or contemporaneous" (see also Merton, [1957] 1973c, pp. 294–295). Moreover, in direct contrast to commercial relations, the greater the "use" of the property by others, the greater the value of the property (see also Price, 1978, pp 80–81). The availability of knowledge claims is the condition for the possibility of a "continued appraisal of work and recognition for work judged well done *by the standards of the time* [which] constitute a mechanism for maintaining the processes of falsification and confirmation of ideas that are required for the cognitive *development* of science" (Merton, 1976, p. 45; emphasis added).

3. These distinctions suggest that the ethos of science (norms) pertains directly to what could be called both methodological (cognitive) and social means, which are simultaneously sub-

sidiary to the goal of science (values). Such categorical distinctions indeed are made at times (for instance, Storer, 1966, p. 76), but I shall not attempt to maintain such a distinction here. That would presuppose a much more explicit and developed theory of the social relations of science than actually is available at this time.

4. The decision to concentrate attention on the normative basis of the social relations of science presumably also rests on a number of more nearly immanent meta-theoretical considerations. For example, as emphasized at various times (Merton, 1948, 1975), a general theory of social action is not a realistic prospect in sociology at the present stage of its cognitive development; thus, more delimited theoretical structures have to be developed, and research strategies have to be adjusted accordingly. The scope of theories is therefore limited by the cognitive state of the discipline. At the same time, one of the central features of Mertonian discourse, as described by Coser (1975, p. 5), might well be relevant: "Merton's world is composed of multiple ambiguities, of conflicting and contradictory demands and requirements that need to be articulated and made accessible by the sociologist."

5. However, I will exclude a more detailed discussion of the possible interrelation between societal (including cognitive) norms and the ethos of science (see also Merton, [1938] 1973d, p. 259; Barber, 1952; Blume, 1974, pp. 45–50), including the views advanced by Downey (1967, p. 253) that "scientism" assures the obsolescence of a number of norms which are part of the ethos of science, or that the autonomy of scientific institutions is in effect the outcome of cognitive progress in science.

6. Often, as I have tried to show elsewhere (Stehr, 1975), these domain assumptions become much more explicit in the writings of authors who attempt to follow Merton in their analysis of science as a social institution; for example, Storer (1966, p. 82) emphasizes the point that there is but *one* science.

7. I refer, of course, to Rickert and Weber's theory of concept formation in the empirical sciences, particularly Weber's notion of idea types as conceptual tools in the social sciences. Such idea types were not conceived simply as descriptions of observable conduct (Rickert, 1902; Weber, 1922; Burger, 1976).

8. The position that Mulkay advances is not, it seems, peculiar for a social scientist to adopt (Gaston, 1978, pp. 163–166) but rather is the result of underlying Kuhnian presuppositions. Gaston, in his critique of Mulkay's view, discusses the norms of

science primarily as they manifest themselves in the social relations of science (the research formulation), whereas Mulkay discusses the ethos of science primarily in terms of cognitive processes and consequences in science. Kuhn's program for paradigm closure (Simmons and Stehr, 1978) and logical empiricism as well as critical rationalism (Hempel, 1965, pp. 3–5; Popper, [1935] 1968, p. 41; Popper, [1962] 1969, p. 156) also is based on a differentiation of science from nonscience; namely, the idea that demarcation must be based on cognitive attributes of systems of discourse. As far as social relations in general are concerned, positivism emphasizes cognitive factors of social action, while functionalism stresses norms, values, motives, and the consequences of social conduct (King, 1971, p. 8; see also Parsons, [1937] 1949, pp. 387, 439–440). The emphasis of cognitive norms to the exclusion of other processes in science, of course, results in an image of a rather "rational" scientific practice. It may therefore fail to do justice to the immense complexity of the social organization of science. That is, the reduction (and retention) of complexity in science, to use Luhmann's terminology, requires social mechanisms other than cognitive norms—for example, reputation operating *both* as a medium of communication (selection) and as motivation (Luhmann, [1968] 1970).

9. In an essay written some thirty years after the original formulation of the ethos of science, Merton [1968] 1973a, pp. 327–328) appears to acknowledge the historicity of the social norms and with it, of course, the possibility that the norms of science are neither entirely autonomous nor unresponsive to basic sociostructural changes in science and society. However, the emphasis, in this essay at least, is on the continuity of certain institutional relations of science, such as the intensity and the degree of competition among scientists.

10. In his essay "The Ambivalence of Scientists," Merton (1976, pp. 41–42) refers to the doubling of ambivalence—scientists' ambivalence toward feeling ambivalent—as a possible explanation of the lack of attention paid to conflicts and modes of conflict in science as conflicts over claims to priority. One of the interesting issues raised by the conception of conflicting norms as a "normal," structurally generated attribute of social organization concerns the problem of deviant conduct under such circumstances. For deviant conduct traditionally is defined as behavior at variance with specific dominant norms typifying social

positions. Given the particular formulation of norms and counter-norms (their scope and range, for instance), deviant conduct might well be denied *a priori.* In the case of science, however, deviant conduct also requires a set of rules signifying both social and cognitive distinctiveness of scientific practice (demarcation criteria). (For a discussion of deviant behavior from both cognitive and social norms in science, see Zuckerman, 1977.)

11. The empirical evidence Mitroff reports refers to research about the moon, which at the time was not intellectually homogeneous (in the sense of a dominant paradigm, for instance). As Mitroff (1974b, p. 594) therefore suggests, "whereas the conventional norms of science are dominant for well-structured problems, the counternorms proposed here appear to be dominant for ill-structured problems."

12. Mulkay's underlying theory of institutions largely rests on a notion of social control (and norms) based on externally imposed sanctions (or sanctioned norms) but not on what may be called "primary" social control. The internalization of norms does not require an "exchange" or reward (Berger and Luckmann, 1966). A critique of such a restrictive conceptualization of social norms also may be found in Luhmann (1969), while Dahrendorf (1964) and Popitz (1961) advocate the restriction of the concept of norm to sanctioned conduct. The somewhat retrogressive perspective argued by Mulkay, at least in the context of his discussion of the status of the norms of science, can also be illustrated with respect to the theoretical status of "intellectual commitments." Mulkay largely abstains from linking commitments to social mechanisms. Social mechanisms may explain the homogeneity (within specific social boundaries), heterogeneity, distribution, and reproduction (or origin and alteration) of intellectual commitments. Such an explanation requires a theoretical analysis that transcends either a cognitive approach or a sociology of scientific ideas and procedures.

13. Using information gathered for a small and heterogeneous sample of university scientists, West (1960, p. 61) reports that neither their rate of publication nor the strength of their motivation as reported by peers appears to be associated with an endorsement of the ethos of science.

14. Following Feyerabend (1975) and Churchman (1971), Mitroff (for example, 1974b, p. 590) subscribes to a historiography of scientific knowledge based on the theory of paradigm proliferation. Subsequently, Merton (1976, p. 59) has criticized Mitroff for

exaggerating the inevitably "subjective" attributes of the production of scientific knowledge to the detriment of its objective aspects and for ignoring the interaction of subjective and objective attributes, which results in a "storybook version of scientific inquiry."

References

Baldamus, W. "Ludwig Fleck and the Development of the Sociology of Science. In *Human Figurations: Essays for Norbert Elias.* Amsterdam: Amsterdams Sociologisch Tijdschrift, 1977.

Barber, B. *Science and the Social Order.* New York: Free Press, 1952.

Barber, B. "The Resistance by Scientists to Scientific Discovery." *Science,* 1961, *134,* 596–602.

Barnes, S. B. "The Comparison of Belief-Systems Anomaly versus Falsehood." *European Journal of Sociology,* 1970, *11,* 3–25.

Barnes, S. B., and Dolby, R. G. A. "The Scientific Ethos: A Deviant Viewpoint." *European Journal of Sociology,* 1970, *11,* 3–25.

Ben-David, J. "Organization, Social Control and Cognitive Change in Science." In J. Ben-David and T. Clark (Eds.), *Culture and Its Creators.* Chicago: University of Chicago Press, 1977.

Berger, P. L., and Luckmann, T. *The Social Construction of Reality. A Treatise in the Sociology of Kowledge.* New York: Doubleday, 1966.

Blume, S. S. *Toward a Political Sociology of Science.* New York: Free Press, 1974.

Boehme, G. "Die Bedeutung von Experimentalregeln für die Wissenschaft" ["The Importance of Experimental Rules for Science"]. *Zeitschrift für Soziologie* [*Journal of Sociology*], 1974, *3,* 5–17.

Boehme, G., van den Daele, W., and Krohn, W. "Alternativen in der Wissenschaft" ["Alternatives in Science"]. *Zeitschrift für Soziologie,* 1972, *1,* 302–316.

Bourdieu, P. "The Specificity of the Scientific Field and the Social Conditions of the Progress of Reason." *Social Science Information,* 1975, *14,* 19–47.

Box, S., and Cotgrove, S. "Scientific Identity, Occupational Selection and Role Strain." *British Journal of Sociology,* 1966, *17,* 20–28.

Buehl, W. L. *Einführung in die Wissenschaftssoziologie.* [*Introduction to the Sociology of Science*]. Munich: Beck, 1974.

Burger, T. *Max Weber's Theory of Concept Formation.* Durham, N.C.: Duke University Press, 1976.

Churchman, C. W. *The Design of Inquiring Systems.* New York: Basic Books, 1971.

Cole, J. R., and Cole, S. *Social Stratification in Science.* Chicago: University of Chicago Press, 1973.

Cole, J. R., and Zuckerman, H. "The Emergence of a Scientific Specialty: The Self-exemplifying Case of the Sociology of Science." In L.A. Coser (Ed.), *The Idea of Social Structure.* New York: Harcourt Brace Jovanovich, 1975.

Coser, L. A. (Ed.). *The Idea of Social Structure.* New York: Harcourt Brace Jovanovich, 1975.

Coser, L. A., and Nisbet, R. "Merton and the Contemporary Mind: An Affectionate Dialogue." In L. A. Coser (Ed.), *The Idea of Social Structure.* New York: Harcourt Brace Jovanovich, 1975.

Cournand, A. F., "The Code of the Scientist and Its Relationship to Ethics." *Science,* 1977, *198,* 699–705.

Cournand, A. F., and Zuckerman, H. A. "The Code of Science." *Studium Generale,* 1970, *23,* 941–962.

Dahrendorf, R. *Homo Sociologicus.* Opladen: Westdeutscher Verlag, 1964.

Deutscher, I. "Public and Private Opinions: Social Situations and Multiple Realities." In S. Z. Nagi and R. G. Corwin (Eds.), *The Social Contexts of Research.* New York: Wiley, 1972.

Downey, K. J. "Sociology and the Modern Scientific Revolution." *Sociological Quarterly,* 1967, *8,* 239–254.

Feyerabend, P. *Against Method.* London: New Left Books, 1975.

Fleck, L. *Entstehung und Entwicklung einer wissenschaftlichen Tatsache: Einführung in die Lehre vom Denkstil und Denkkollektiv* [*Origin and Evolution of a Scientific Fact: Introduction to the Theory of Styles and Collective Carriers of Thought*]. Basel, Switzerland: Schwab, 1935.

Gaston, J. "Soziale Organisation, Kodifizierung des Wissens, und das Belohnungssystem der Wissenschaft" ["Social Organization, Codification of Knowledge, and the Reward System of Science"]. In N. Stehr and R. König (Eds.), *Wissenschaftssoziologie* [Sociology of Science]. Opladen: Westdeutscher Verlag, 1975.

Gaston, J. *The Reward System in British and American Science.* New York: Wiley-Interscience, 1978.

Gouldner, A. W. *The Dialectic of Ideology and Technology: The Origins, Grammar, and Future of Ideology.* New York: Seabury Press, 1976.

Gutting, G. "Conceptual Structures and Scientific Change." *Studies in History and Philosophy of Science,* 1973, *4,* 209–230.

Hagstrom, W. O. *The Scientific Community.* New York: Basic Books, 1965.

Hempel, C. *Aspects of Scientific Explanation.* New York: Free Press, 1965.

King, M. D. "Reason, Tradition, and the Progressiveness of Science." *History and Theory,* 1971, *10,* 3–32.

Knorr, K. "Producing and Reproducing Knowledge: Descriptive or Constructive? Toward a Model of Research Production." *Social Science Information,* 1977, *16,* 669–696.

Kornhauser, W. (with the assistance of W. O. Hagstrom). *Scientists in Industry.* Berkeley: University of California Press, 1963.

Kowalewski, A. "Bureaucratic Trends in the Organization and Institutionalisation of Scientific Activity." In R. Whitley (Ed.), *Social Processes of Scientific Development.* London: Routledge & Kegan Paul, 1974.

Krohn, R. G. "The Institutional Location of the Scientist and His Scientific Values." *IRE Transactions on Engineering Management,* 1961, *EM-8,* 133–138.

Kuhn, T. S. *The Structure of Scientific Revolutions.* Chicago: University of Chicago Press, 1962.

Kuhn, T. S. *The Essential Tension: Selected Studies in Scientific Tradition and Change.* Chicago: University of Chicago Press, 1977.

Lammers, C. "Mono- and Poly-paradigmatic developments in Natural and Social Sciences." In R. Whitley (Ed.), *Social Processes of Scientific Development.* London: Routledge & Kegan Paul, 1974.

Lautmann, R. *Wert und Norm: Begriffsanalysen für die Soziologie* [*Value and Norm: Concepts of Analysis for Sociology*]. Opladen: Westdeutscher Verlag, 1971.

Luhmann, N. "Selbststeuerung der Wissenschaft" ["Feedback System of Science"] [1968] In *Soziologische Aufklärung: Aufsätze zur Theorie sozialer Systeme.* [*Sociological Enlightenment: Essays on the Theory of Social Systems*]. Opladen: Westdeutscher Verlag, 1970.

Luhmann, N. "Normen in soziologischer Perspektive" ["Norms in the Sociological Perspective"] *Soziale Welt* [*Social World*], 1969, *20,* 28–48.

Marcson, S. *The Scientist in American Industry.* New York: Harper & Row, 1960.

Merton, R. K. "On the Position of Sociological Theory." *American Sociological Review,* 1948, *13,* 164–168.
Merton, R. K. "Behavior Patterns of Scientists" [1968]. In *The Sociology of Science.* Chicago: University of Chicago Press, 1973a.
Merton, R. K. "The Normative Structure of Science" [1942]. In *The Sociology of Science.* Chicago: University of Chicago Press, 1973b.
Merton, R. K., "Priorities in Scientific Discovery" [1957]. In *The Sociology of Science.* Chicago: University of Chicago Press, 1973c.
Merton, R. K. "Science and the Social Order" [1938]. In *The Sociology of Science.* Chicago: University of Chicago Press, 1973d.
Merton, R. K. *The Sociology of Science.* Chicago: University of Chicago Press, 1973e.
Merton, R. K. "Structural Analysis in Sociology." In P. Blau (Ed.), *Approaches to the Study of Social Structure.* New York: Free Press, 1975.
Merton, R. K. *Sociological Ambivalence and Other Essays.* New York: Free Press, 1976.
Merton, R. K. "The Sociology of Science: An Episodic Memoir." In R. K. Merton and J. Gaston (Eds.), *The Sociology of Science in Europe.* Carbondale: Southern Illinois University Press, 1977.
Mitroff, I. I. *The Subjective Side of Science: A Philosophical Inquiry into the Psychology of the Apollo Moon Scientists.* Amsterdam: Elsevier, 1974a.
Mitroff, I. I. "Norms and Counter-Norms in a Select Group of the Apollo Moon Scientists: A Case Study of the Ambivalence of Scientists." *American Sociological Review,* 1974b, *39,* 579–595.
Mok, A., and Westerdiep, A. "Societal Influences on the Choice of Research Topics of Biologists." In R. Whitley (Ed.), *Social Processes of Scientific Development.* London: Routledge & Kegan Paul, 1974.
Mulkay, M. J. "Some Aspects of Cultural Growth in the Natural Sciences." *Social Research,* 1969, *36,* 22–52.
Mulkay, M. J. "Conformity and Innovation in Science." In P. Halmos (Ed.), *The Sociology of Science.* Sociological Review Monograph No. 18. Keele, England: University of Keele, 1972a.
Mulkay, M. J. *The Social Process of Innovation: A Study in the Sociology of Science.* London: Macmillan, 1972b.
Mulkay, M. J. "Drei Modelle der Wissenschaftsentwicklung" ["Three Models of the Evolution of Science"]. In N. Stehr and R.

König (Eds.), *Wissenschaftssoziologie* [*Sociology of Science*]. Opladen: Westdeutscher Verlag, 1975.

Mulkay, M. J. "Norms and Ideology in Science." *Social Science Information,* 1976, *15,* 627–656.

Parsons, T. "The Professions and Social Structure." *Social Forces,* 1939, *17,* 457–467.

Parsons, T. *The Structure of Social Action.* Vol. 1. New York: Free Press, [1937] 1949.

Parsons, T. *The Social System.* New York: Free Press, 1951.

Parsons, T. "On building Social Systems Theory: A Personal History." *Daedalus,* 1970, *99,* 826–881.

Popitz, H. "Soziale Normen" ["Social Norms"]. *European Journal of Sociology,* 1961, *2,* 185–198.

Popper, K. R. *The Logic of Scientific Discovery.* New York: Basic Books, 1935 (enlarged ed., 1959). London: Hutchinson, 1968.

Popper, K. R. *Conjectures and Refutations.* London: Routledge & Kegan Paul, [1962] 1969.

Price, D. de S. "Toward a Model for Science Indicators." In Y. Elkana and others (Eds.), *Toward a Metric of Science: The Advent of Science Indicators.* New York: Wiley-Interscience, 1978.

Ravetz, J. R. *Scientific Knowledge and Its Social Problems.* New York: Oxford University Press, 1971.

Rickert, H. *Die Grenzen der naturwissenschaftlichen Begriffsbildung* [*The Limits of Concept Formation in Natural Science*]. Tübingen: Mohr, 1902.

Rothman, R. A. "A Dissenting View on the Scientific Ethos." *British Journal of Sociology,* 1972, *23,* 102–108.

Scherhorn, G. "Der Wettbewerb in der Erfahrungswissenschaft" ["Competition in Empirical Science"] *Hamburger Jahrbuch für Wirtschafts- und Gesellschaftspolitik* [*Hamburg Yearbook of Economic and Social Policy*], 1968, *14,* 63–86.

Shephard, H. A. "The Value System of a University Research Group." *American Sociological Review,* 1954, *19,* 456–462.

Simmons, A., and Stehr, N. "Language and the Growth of Knowledge in Sociology." Unpublished paper, 1978.

Sklair, L. *Organized Knowledge. A Sociological View of Science and Technology.* St. Albans, England: Paladin Books, 1973.

Spiegel-Rösing, I. *Wissenschaftsentwicklung und Wissenschafts-*

steuerung: Einführung und Material zur Wissenschaftsforschung [*Evolution and the Control of Science: An Introduction and Materials on Science Research*]. Frankfurt am Main: Athenäum, 1973.

Spinner, H. F. "Thesen zum Thema Reichweite und Relezanz der Wissenschaftstheorie für die Einzelwissenschaften—Analytische Philosophie Versus Marxismus" ["The Range and Relevance of Philosophy of Science Theory for the Substantive Sciences—Analytic Philosophy Versus Marxism"]. In K. -H. Braun and K. Holzkamp (Eds.), *Kritische Psychologie* [*Critical Psychology*]. Vol 2. Cologne: Pahl-Rugenstein, 1977.

Stehr, N. "Zur Soziologie der Wissenschaftssoziologie" ["On the Sociology of the Sociology of Science"]. In N. Stehr and R. König (Eds.), *Wissenschaftssoziologie* [*Sociology of Science*]. Opladen: Westdeutscher Verlag, 1975.

Storer, N. W. *The Social System of Science.* New York: Holt, Rinehart and Winston, 1966.

Storer, N. W. "Introduction." In R. K. Merton, *The Sociology of Science: Theoretical and Empirical Investigations.* Chicago: University of Chicago Press, 1973.

Useem, M. "Scientific Normative Orientations and Research Methodologies: A Comparative Study of European Scientists and Engineers at Euratom." Unpublished report, 1968.

Weber, M. *Gesammelte Aufsätze zur Wissenschaftslehre* [*Collected Essays on the Theory of Knowledge*]. Tübingen: Mohr, 1922.

Weingart, P. "Wissenschaftsforschung und Wissenschaftssoziologie ["Research on Science and Sociology of Science"]. In P. Weingart (Ed.), *Wissenschaftssoziologie. 1: Wissenschaftliche Entwicklung als sozialer Prozess* [*Sociology of Science: The Evolution of Science as a Social Process*]. Frankfurt am Main: Athenäum, 1972.

Weingart, P. "On a Sociological Theory of Scientific Change." In R. Whitley (Ed.), *Social Processes of Scientific Development.* London: Routledge & Kegan Paul, 1974.

Weingart, P. *Wissensproduktion und Soziale Struktur* [*The Production of Knowledge and Social Structure*]. Frankfurt am Main: Suhrkamp, 1976.

West, S. S. "The Ideology of Academic Scientists." *IRE Transactions on Engineering Management,* 1960, *EM-1,* 54–62.

Whitley, R. D. "Black Boxism and the Sociology of Science." In P. Halmos (Ed.), *The Sociology of Science.* Sociological Review Monograph No. 18. Keele, England: University of Keele, 1972.

Wunderlich, R. "The Scientific Ethos: A Clarification." *British Journal of Sociology,* 1974, *25,* 373–377.

Young, R. "The Historiographic and Ideological Contexts of the Nineteenth-Century Debate on Man's Place in Nature." In M. Teich and R. Young (Eds.), *Changing Perspectives in the History of Science.* London: Heinemann, 1973.

Ziman, J. *Public Knowledge.* Cambridge, England: Cambridge University Press, 1968.

Zuckerman, H. "Deviant Behavior and Social Control in Science." In E. Sagarin (Ed.), *Deviance and Social Change.* Beverly Hills: Sage, 1977.

Zuckerman, H., and Merton, R. K., "Patterns of Evaluation in Science: Institutionalization, Structure and Functions of the Referee System." *Minerva,* 1971, *9,* 66–100.

10

Joseph Ben-David

Emergence of National Traditions in the Sociology of Science

The United States and Great Britain

This chapter describes the emergence of two different traditions in the sociology of science in the United States and Britain, the differences resulting from the different backgrounds and professional functions of the sociologists of science in the two countries. (For histories and surveys of the development of sociology of science in general, see Cole and Zuckerman, 1975; Stehr, 1975; Merton and Gaston, 1977; Mulkay, 1977a, 1977b.)

Sociology of Science in the United States

Sociology of science as a distinct specialty emerged in the United States in the 1950s as a result of the work of Robert Merton

Note: This research was supported by a grant from the Ford Foundation. A first draft was written while I was a member of the Institute for Advanced Study at Princeton and was read at a joint meeting of the Society for the Social Studies of Science and the Research Committee on Sociology of Science of the International Sociological Association at Cornell University, Nov. 1976.

and his students. Others before him, including several sociologists, had investigated the social aspects of science; but only Merton and his group made a conscious effort to establish a definition of the area, a conceptual framework, and a program of research, and they were the first to make a conscious effort at obtaining recognition for the field as a branch of sociology.

Science and the Social Order (1952), by one of the earliest students of Merton, Bernard Barber, was the first codification of existing knowledge in the field. As is evident from Robert Merton's "Foreword" to the book, which deals with the reasons for the neglect of sociology of science, the purpose of the book was to show that there was enough knowledge and theoretical importance in the field to warrant its recognition as a sociological specialty. The book had great success and became a standard text and reference for many years. It is still an unparalleled example of a comprehensive and systematic outline of the field and a basic document of the program to make sociology of science a recognized specialty. The theoretical core of this program was the description of science as a social institution with a normative structure and a reward system of its own. This "structural-functional" institutional approach to science developed in comparative and historical macrosociology and was useful in the interpretation of differences between cultures, value change within cultures, and the congruency or incongruency between the norms of different institutions in society. (The classic text of this institutional approach to sociology is Davis, 1949.)

Much of the research in the sociology of science before the 1960s dealt with such institutional problems. The best known of these was Merton's discovery of the congruence between Puritanism and science (Merton, [1938a] 1970), which contradicted the belief about an inherent conflict between religion and science. This was followed by investigation of the problem of how science, a universalistic and—in principle—skeptical enterprise, could survive under totalitarian regimes opposed to such norms of behavior (Merton, [1938b] 1973c, [1942], 1973a) and by investigations of the effect on industrial research of the contradiction between the scientific norms of altruism and "communism" and the requirements of secrecy and profitability in industry (Marcson, 1960; Kornhauser, 1962).

Apart from Merton's early study of Puritanism and science, these institutional studies had limited influence. The institutional approach is best suited to the treatment of comparative historical material, but the way sociology developed in the 1950s provided no incentive for the acquisition of historical knowledge. The general expectation was that sociology would follow the example of economics and psychology and adopt quantitative techniques. Few promising graduate students in the 1950s or 1960s in the United States were willing to write a thesis that was not based on quantitative survey research. The alternative of writing theses based on historical-comparative material would have required a kind of erudition that sociology students did not possess; and, in view of the prevailing quantitative trend, they had no compelling reason to acquire such erudition.

If sociology of science was to become a recognized specialty, it had to adopt quantitative techniques (which in the 1950s and early 1960s meant mainly survey research), but the existing (structural-functional) institutional approach was not suited for these. This difficulty was overcome with Merton's discovery of the problematic nature of the reward system in science. Although the crucial paper "Priorities in Scientific Discovery" (1957) was a historical paper written in the tradition of classical qualitative structural-functional analysis, it opened the way to quantitative studies in the sociology of science. Merton's explanation of the apparent incongruency between the selfishly petty behavior of scientists in priority disputes and the scientific norm of "communism" (that is, that scientists might be reluctant to share their results with everyone because they might thereby be deprived of the recognition due to them) suggested that the relationship between the allocation of rewards in science and the behavior expected of scientists was a matter requiring careful study and that the study of competition and stratification could be of central importance in understanding scientific behavior. Competition, allocation of rewards, and stratification could be studied quantitatively, and their quantitative study in science created an opportunity to line up sociology of science alongside general studies of stratification, which constituted one of the central fields of sociological research.

This development of the sociology of science toward articulation with theories of stratification was only a potential in 1957, when Merton's "Priorities" paper was published. The potential was exploited in the 1960s, when Warren Hagstrom, a graduate student at Berkeley, and William Kornhauser, a member of the sociology department, decided to go into the field. In the first result of this collaboration (Kornhauser, 1962), the new theoretical possibilities were not fully exploited. They were exploited, however, in Hagstrom's *The Scientific Community* (1965). Although the techniques that Hagstrom employed were rather rudimentary, he made a systematic effort to measure competition, communication, and recognition.

The historical significance of the book was that it was written not by a student of Merton but by a student at an outstanding department where sociology of science had not been previously represented. This was a sign that the specialty was beginning to be recognized as one in which discoveries of theoretical importance for sociology as a whole could be made.

Indeed, the book was a harbinger of a takeoff in the sociology of science. Several of the most promising doctoral students at Columbia in the 1960s—such as Jonathan Cole, Stephen Cole, Diana Crane, and Harriet Zuckerman—chose sociology of science as their main field of research and established a considerable reputation in sociology in general by working on the themes outlined in Merton's "Priorities" paper and Hagstrom's *Scientific Community*. They have since been joined by others.

This success was due not only to the theoretical potentialities of Merton's ideal but also to other influences in the 1960s and the 1970s. One was the work of Derek de Solla Price, which began in the early 1950s and became widely known in the 1960s through two books, *Science Since Babylon* (1961) and *Little Science, Big Science* (1963). Price was interested in measuring the growth of science and discovering its immanent regularities. His interest in social conditions was limited to the restraints imposed on this growth by the facts that only a small fraction of mankind is capable of cultivating science, and that there are rather narrow limits to the capability of people to transmit and absorb information. Models of the rise and decline of special fields were developed by Holton (1962) and by Price (1963, pp. 22–23). Both works used publications and citations in order to evaluate contributions and analyze the development of specialties (see Price, 1963, pp. 62–91), thereby providing imaginative examples and suggestions for the quantitative treatment of sociological problems in science. *Science Citation Index,* published since 1963, has been an invaluable tool for this work. Scientific recognition and the flow of communication between scientists could now be measured effectively, and on an unprecedented scale, through citations. Thus, by the mid 1960s all the theoretical ingredients and technical tools required for quantitative studies of stratification and reward in science were in existence and recognized by sociologists. Soon thereafter methods for their effective use were developed by Jonathan and Stephen Cole. This availability explains the rapid rise of systematic and coherent work in this field (Cole and Cole, 1973; Gaston, 1973, 1978).

During the 1960s and the 1970s, numerous studies of scientific growth, particularly the growth of particular fields and specialties, appeared (Ben-David, 1960; Ben-David and Collins, 1966; Fisher, 1966; Clark, 1968, 1973; Crane, 1969, 1972; Krantz, 1971; Crawford, 1971; Mullins, 1972, 1973; Griffith and Mullins, 1972; Griffith and others, 1974; Small and Griffith, 1974; Cole and Zuckerman, 1975; Breiger, 1976; Chubin, 1976; Mullins and others, 1977). These studies have not had such a homogeneous theoretical focus as those of reward and stratification. Some of them are conceptually related to the latter, since they explain differences in the growth of science in different countries as the result of differences in prestige and standing between institutions; similarly, they explain the rise of certain new fields as the result of mobility of scientists from field to field, motivated by institutionally determined opportunities. The rise of interest in such studies was greatly influenced by Price's suggestions for the quantitative study of specialty networks, called *invisible colleges* (see Crane, 1972), by the *Science Citation Index;* and by Thomas Kuhn's *The Structure of Scientific Revolutions* (1962). Kuhn's influence was less specific than that of Merton and Price, because his ideas were not readily translatable into empirical research and their sociological implications were not sufficiently clear (Barber, 1963). But no other book has painted such a vivid and sociologically suggestive picture of the scientific community and made such a consistent attempt to describe the rise and decline of scientific traditions ("paradigms") as a combination of intellectual and social processes. His idea that science advances through revolutions aroused great interest in the investigation of scientific discoveries, especially those that could be described as revolutionary. The general influence of Kuhn's ideas and the use of quantitative techniques lent to this line of research a degree of unity, although nothing like the coherence prevailing in the investigations of stratification and rewards.

These two relatively coherent lines of research are not the whole story of American sociology of science. The study of scientific organizations from the point of view of research management has been a much more continuous tradition, which began in the early 1950s and is still continuing (Shepard, 1956; Kaplan, 1960, 1964; Glaser, 1964; Allen, 1966; Gordon and Marquis, 1966; Pelz and Andrews, 1966; Allen and Cohen, 1969). In the 1950s and early 1960s, these studies were closely integrated with the rest of the sociology of science. Since then, however, these investigations have become absorbed in the general area of management studies and are now largely confined to departments of business and public administration; consequently, there has been little contact between investigators of these problems and the rest of the sociologists of science.

The more traditional structural-functional analysis (Storer, 1966) and comparative-historical investigations of the institutions of science (Shils, 1970; Ben-David, 1971; Clark, 1973) also have continued. But the reward and stratification studies were the gate through which the majority of graduate students who eventually made contributions entered the field. This particular problem area produced one of the most continuous, clearly formulated, consciously pursued, and technically advanced research programs in sociology. The existence of such a program in the field also raised the salience and attractiveness (for graduate students) of the other lines of inquiry in sociology of science. The fact that these other lines of inquiry did not constitute clearly formulated programs, and were only loosely coordinated conceptually or methodologically, did not really matter, since much research in sociology is of this loosely coordinated kind, and the existence of even one program is sufficient to lift a field above many others.

The emergence of quantitative studies of scientific specialties in the late 1960s added a new line of investigation to the field. Although it had no clear theory, it had a core of common ideas on how scientific innovations occur; above all, it used quantitative techniques for network analysis. Network analysis was not directly related to the reward and stratification research tradition and the normative structural-functional assumptions that gave rise to the reward and stratification tradition. There was no inconsistency between the two lines of research, but a researcher could pursue one of them without paying attention to the other. The emergence of this new line of research made it possible for researchers to avoid the controversy that surrounded the structural-functional approach.

Sociology of Science in Britain

In Great Britain, sociology of science is much younger and has had a history very different from that in the United States. No British sociologist was interested in the field before the 1960s. A few British scientists, in particular J. D. Bernal (1939), Joseph Needham (1931), and Michael Polanyi (1951), made important contributions to the field during the 1930s and 1940s, but they did not identify themselves as sociologists and did not train students or start programs of research (except Joseph Needham in the history of Chinese science). However, this tradition of prestigious scientists paying serious attention to the social aspects of science created a favorable background for the eventual recognition of social studies of science as an academic field. Therefore, the rise of public interest in the development and social uses of science during the 1950s and 1960s led to the establishment of science units (at Sussex and Edinburgh) and programs for developing courses of study and research in this field (at Manchester). The salience of science and its public discussion attracted to the general field economists, sociologists, philosophers, historians, and people trained in science and engineering. They had neither central leadership nor a common program; but they realized that science had become a socially important phenomenon in their lifetime, and they wanted to understand this development. In contrast to the United States, where this interest was channeled into a preexisting tradition cultivated in graduate departments of sociology, British students had no local tradition and—especially those working in interdisciplinary units—were exposed to a variety of disciplinary backgrounds. Thus, although most economists, historians, philosophers, and sociologists of science are as distinguishable from each other in Britain as elsewhere, the interdisciplinary framework enabled, and perhaps even prompted, a small number of people trained in science of other fields outside sociology—such as B. Barnes, D. Bloor, and R. G. A. Dolby—to become sociologists or to participate in the internal debate on sociological theory.[1]

These British sociologists of science came to the scene in the late 1960s, at a point when the American tradition was beginning to change from overwhelming interest in the reward system to increasing interest in the sociology of scientific specialties. Moreover, even in the United States, the professional definition of the field was much looser than before because of the impact of the work of two non-sociologists, Derek de Solla Price and Thomas S. Kuhn, and the increasingly vocal attacks on structural-functional analysis. The difference in the background and organizational and institutional conditions of work between the American and British group was most obvious in the way the two groups reacted to this situation, and particularly in the way they received the ideas of Kuhn. As has been shown, in the United States this influence was filtered through a strong disciplinary background in sociology. Kuhn's ideas about the developmental phases of scientific knowledge aroused the greatest interest, because these had the most obvious sociological contents and promised to be capable of empirical verification. Much less attention was paid to Kuhn's philosophical relativism and its implications for a sociology of knowledge, partly because Kuhn himself was not too clear about his own relativism (the "commensurability" of "paradigms") and partly because professional sociologists were acquainted with Merton's ([1945] 1973b) essay "Paradigm for the Sociology of Knowledge" and were aware of the immense difficulty involved in investigations of the sociology of knowledge. Thus, until about 1970—that is, for about a decade—American sociologists of science did not believe that Kuhn's views competed with those of Merton. They used and quoted both, for different purposes.

In Britain, the reception of Kuhn was very different. Perhaps because of their background in other disciplines, some British sociologists of science could not appreciate the theoretical importance of interpreting science in terms that had systematic sociological meaning. And those who did were usually critical of structural-functional theory. This, of course, was true also of many Americans. But in the United States there was a tendency to avoid purely theoretical debate (at least in the sociology of science), while in Britain—for reasons to be explained below—many sociologists made their careers through debate and criticism. As a result, the implantation of American sociology of science in Britain was accompanied by a reordering of the cognitive structure of the field. Instead of viewing the ideas of Merton, Kuhn, and Price as so many attempts at conceptualizing the complexities of the scientific community, to be used as inputs in an effort to unravel the structure and function of that community by piecemeal empirical research, British sociologists analyzed these ideas philosophically, for their internal consistency and their logical compatibility with one another. Thus, as one of their main innovations, the British sociologists set up two opposing models: the Mertonian "model" of a general scientific community acting according to relatively stable norms and the Kuhnian "model" of scientific communities changing their views and rules through periodic revolutions. Having set up these models as mutually exclusive (which, as has been pointed out, they were not), they criticized the Mertonian model as incongruent with some (not very systematic) observations on how scientists behave and opted for the Kuhnian model (Mulkay, 1969; Barnes and Dolby, 1970; Dolby, 1971; King, 1971; Martins, 1971). They did so not because Kuhn's model was more congruent with observations (which they did not try to check) but because it dealt with scientific communities as defined by their members' intellectual and scientific concerns, which for everyone except professional sociologists was a much more interesting point of view than the analysis of the norms and the reward system of science. (See the disappointed reaction of Charles Gillispie to Merton's "Priorities" paper, described by Cole and Zuckerman, 1975, p. 157.)

The adoption of Kuhn's ideas as a starting point for a program of research was not confined to Britain. As I have pointed out, there was such a trend in the United States as well. But, as long as both Merton's and Kuhn's ideas were treated as hypotheses about different aspects of scientific behavior to be empirically investigated, the possibility of contradictions between some of the implications of these ideas was of little interest. However, when the ideas were approached from a philosophical point of view, then the difference between Kuhn's qualified relativism and Merton's emphasis on relatively stable institutionalized norms of scientific behavior seemed interesting to explore.

This preponderance of philosophical interest among British sociologists is partly a result of the role of sociology in Britain. British academic sociology is primarily undergraduate sociology. The aim of instruction is not to train research workers but to teach people to think, talk, and write about social issues in a clear, coherent, and effective fashion. Theory, or actually a kind of social philosophy, is, therefore, a central rather than a peripheral aspect of sociological study. Sociologists are taught to confront alternative views of society, or aspects of it; to present these views so that they appear internally coherent and mutually exclusive; and to "demolish" some or all of the views by argument. Writing on such subjects—which in the United States would be considered a marginal contribution to scholarship, qualifying one as a college teacher but hardly for appointment in a graduate department—is, apparently, a highly valued activity in Britain. Thus, about half of the sociology of science literature in Britain (an estimate based on classification and count of publications by British authors quoted in Mulkay, 1977a) consists of this kind of "theoretical" writing; that is, critical summaries and confrontations of views on the social aspects of science, in particular of Kuhn *versus* Merton.

Another reason for this interest in Kuhn's relativism is the fact that some of the people in the field, as has been pointed out, are not sociologists by training but scientists and/or philosophers. Several of them do not work in sociology departments but in special interdisciplinary units, charged with giving courses on and generating interest (especially among science and engineering students) in the social aspects of science and, occasionally, with training future (or present) science administrators in the same area. For these types of teachers and/or students, the prospect of learning from science something about the working of a "fair," consensual social reward system—which is one of the most interesting aspects of science for sociologists—is not only uninteresting but actually disturbing. If members of the scientific community act according to the generally accepted norms of science, then sociology adds nothing to what they were taught (explicitly or implicitly) about science. Sociology becomes interesting only if it can show that actually there is no consensus in science and that decisions about what is at any moment accepted as scientific "truth" are arrived at through a process of conflict of interests, power struggle, and "negotiation," as in many other fields of behavior.[2] One can "really" understand science, then, only through disclosure of the social processes that create it. Therefore, it was important to show that the social aspects of science are not confined to the determination of the place and effects of science in society and to social influences on the institutionalization of science and scientific specialties, but that the very substance, or cognitive content, of science is also socially determined. The British thus became deeply interested in the sociology of scientific knowledge (Mulkay, 1977a, p. 245) and impatient with the sociology of science tradition concerned with the institutional aspects of science. Because of their lack of sociological background, some of the British proponents were unaware of the less than glorious history of attempts at a sociology of knowledge, and of the extreme difficulties involved in such efforts.[3]

The two national traditions in the sociology of science are not sharply demarcated from each other, and in some areas, such as studies of research management or to some extent studies of the rise and decline of specialties, the traditions overlap considerably. Nevertheless, the two traditions are clearly distinguishable. American research has emphasized the description of general norms of scientific behavior; the exploration of the reward system and stratification in science, and the formation of consensus in the evaluation of scientific work and merit. Studies on these themes have usually been conducted in a framework of structural-functional assumptions, according to which science in general is a well-demarcated, institutionalized activity. Few American sociologists have tackled problems of sociology of knowledge in science. There has been a great preference in American studies for the use of quantitative techniques.

British sociologists of science have been much more concerned with critical evaluation of existing research and with attempts (mainly theoretical) at creating a sociology of scientific knowledge. On the whole, they have been critical of structural-functional analysis, and their view of science has tended to be much more relativistic than in the United States. According to them, scientific norms and "truths" are changing from field to field and time to time, more under the impact of "negotiations" between opposing interests than of new discoveries. Therefore, they are much more interested in scientific conflicts than in consensus. They have concentrated on the sociology of particular fields rather than on science in general and have only employed quantitative techniques in rare cases.

As has been pointed out, these differences are related to differences in the intellectual background and professional functions of the two groups. The Americans are professionally trained sociologists working at graduate departments. They have intellectual contacts with historians and philosophers of science who have sociological interests, but professionally they are clearly separated from them. In Britain the definition of who is a sociologist is much looser. Many of the British were trained in science and philosophy and became sociologists after the completion of their formal training. British sociologists of science teach mainly undergraduates, and are frequently attached to interdisciplinary units.

Historically, the American tradition preceded the British one, and part of the difference between them is probably due to the fact that the British tradition emerged at a time when the American tradition was in a state of transition and when structural functionalism was also being strongly criticized in the United States. Thus some American sociologists from the younger generation (and a number of American historians with a strong interest in sociology) are sympathetic to the British views, but theirs is only a philosophical sympathy. In their research work they follow the local, rather than the British, style. Therefore, the difference between the two traditions is not merely a generational one, and there remains a difference even if one compares scholars of the same generation from the two countries.

The preceding explanation of the differences between the styles of work, preferences in problem choice, and philosophical views of science does not imply any evaluation of the relative merits of the two national traditions. Although they are often presented, especially in Britain, as mutually exclusive, they are actually parallel or even complementary traditions. It is not less interesting to investigate institutional characteristics (norms, rewards, formal and informal social structures) common to all scientific fields than to investigate social characteristics peculiar to certain fields, or events in science; it is as legitimate to ask questions about consensus as about conflicts in science; and it is difficult to tell whether it is more important to concentrate on how the intellectual characteristics common to all the sciences influence the behavior of scientists or on how the social characteristics of scientists affect the contents of their work.

This chapter illustrates the complementarity rather than mutual exclusiveness of these questions. It deals with scientific conflict and the effects of the social characteristics of scientists on the contents of their work—questions typical to the British tradition. However, what is said here cannot be construed as supporting the philosophical views preferred in Britain, namely that scientific views are determined by social ("external") conditions, rather than by the internal logic of scientific traditions and inherent characteristics of the phenomenal world, and that permanent conflict, rather than the formation of consensus, is the typical process of science. The chapter only shows that in this particular case the problems and approaches preferred by members of two groups of sociologists were socially determined; it says nothing about the origins of those problems and approaches. If we were to inquire into origins, we would find that they are internally determined.

The problems and approaches from which members of both groups made their choices were derived from ideas available to them in the sociological and philosophical traditions of the 1960s. These were the structural-functional approach as developed by Talcott Parsons and others (including Robert Merton); the attempts at correcting, modifying, or replacing that approach which emerged in the 1950s in the wake of the comprehensive effort to develop instructional-functional theory; and survey research and analysis as initiated by Samuel Stouffer and Paul Lazarsfeld. In philosophy, Karl Popper provided a concept of science that aptly characterized what has been common to all scientific fields since the seventeenth century and therefore was excellently suited for the study of science as a social institution. But for the same reason it also presented an obvious challenge to the next generation of philosophers of science—such as P. K. Feyerabend, T. S. Kuhn, Imre Lakatos, and Stephen Toulmin—who found Popper's concept of science insufficient for the analysis of scientific change and the differences between scientific change and the differences between scientific fields. Their attempts to create a more historicist view of science led to a revival of interest in the sociology of knowledge of Durkheim and Mannheim. Without this background of the internal history of sociological and philosophical ideas, sociological interpretations would make little sense, since they could not explain how the interests of these groups actually led to the creation of the particular ideas characteristic to them. In other words, without the existence of a common background that has an "internal" logical structure and determines the available alternatives, the choices made by the two groups are incomprehensible. A historian concentrating on the emergence of present issues and approaches in the sociology of science (and uninterested in the question of why some people, or groups, chose some rather than other issues and approaches) could easily depict the present state of the field as characterized by internalist doctrine. Such a presentation would also be a consensual one and would assume the existence of common norms, since it would place controversies into a commonly agreed-on framework of problems and would discuss any divergences of view as transitory states to be eliminated by further investigation conducted according to recognized norms.

Of course, both the externalist-conflictual and internalist-consensual presentation would be partial ones. In this field, as in other fields of science, we have two kinds of processes: One is the ongoing emergence of groups devoted to the exploration of a particular issue or of a particular set of hypotheses. The emergence and composition of these groups are likely to be influenced by social conditions, although such groups will usually work on problems derived from existing traditions or combination of traditions (Ben-David, 1960; Ben-David and Collins, 1966). The relationship to other groups working in the same or related areas will often be determined by competition for scarce resources and rewards (such as recognition, appointments to positions, research grants, or honors), which is the second process. Especially in nonexpanding systems, this competition may lead to conflicts, such as attempts by the well-established groups to suppress new ones, or attempts by the new groups to overthrow well-established ones. Although in the large majority of cases the conflict is over resources and rewards, and not about mutually exclusive explanations of the same phenomena, such groups will often try to present their views as a contradiction and refutation of those of competing groups and thus justify their claims for withholding rewards and resources from the latter.[4] Such conflicts may vary in intensity but are usually of short duration. Typically, they are resolved by recognition of the new field as a new specialty—that is, by growing differentiation in the scientific division of labor rather than by revolutionary takeovers or counterrevolutionary suppression (see for example Zloczower, 1960). In other words, conflicts are resolved by the formation of a new consensus.

Thus there is in science a long-term process that counteracts short-term conflicts. This long-term process determines the selection of ideas, problems, and solutions for incorporation into the traditions of given scientific fields. It involves several groups and generations, some of whom are neutral toward the original conflict of interests. Their attitudes toward the contribution of the competing groups, therefore, will be determined by the utility of those contributions for the rational explanation of the phenomena in question. Of course, neither the norms of selection nor the definition of the phenomena will be quite stable, but, nevertheless, when people act in contexts related to this selection process (such as refereeing papers; reviewing published work; awarding grants, positions, or prizes; or scanning the literature of adjacent fields for information and ideas useful for their own research), they will tend to adopt much more universalistic criteria and will try to act more dispassionately than when competing for resources and recognition with other groups (Ben-David, 1977).

The existence of such a process of selection is also observable in the brief history of the two national traditions described in this chapter. Evaluation of individual contributions is often taking place according to criteria common to both traditions, and evaluation eventually influences practice. Thus reviews of literature have been critical of the absence of empirical studies of the relationship between social processes and cognitive development in science (Mulkay, 1977a, p. 136). It seems that such criticism is having an effect on research, and current work is increasingly turning away from programmatic statements and toward empirical investigations. These investigations will be judged on the basis of their originality, internal consistency, and cogency of their evidence—an evaluation process that will probably lead to a gradual convergence of judgments about the contributions of the two schools.

However, we must keep in mind that the relationship between short-term and long-term processes in science is a purely historical one. There is no law of nature, or progress, to ensure that the sectarian spirit of conflict that arises in some scientific groups will be inevitably subjected to a selection process according to the traditional norms of science. It took the efforts and inventiveness of many generations to establish an institutional mechanism in natural science that has worked more or less as here outlined since the seventeenth century (without, however, eliminating quack medicine, magic or astrology, or—under appropriate conditions—phenomena such as Lysenkoism). In the social sciences the mechanism works much less smoothly, and attempts at sectarian closure of schools is an everyday occurrence on the social scientific scene, as exemplified by psychoanalytic, Marxist, and similar groups. Competition capable of transcending conflict of interests between particular groups is no more automatically ensured in the realm of scientific ideas than in the realm of politics and economics. There will always be groups that prefer monopoly and some that will only have a chance when granted monopolistic privileges, and such groups will do whatever they can do to obtain it.

Notes

1. There have been no comparable cases in the United States. Psychologists like Griffith (Griffith and Mullins, 1972; Griffith and others, 1974) and Krantz (1970, 1971), who did sociological work in this field, have maintained their professional identity. And none of the American historians and philosophers—some of whom have as strong an interest in the sociology of science as their British colleagues—published in sociological publications or intervened in sociological controversies.

2. For a testimony that philosophers and historians of science are particularly interested in what can be learned from social science about the nonconsensual aspects of science, see Kuhn (1970, p. viii). Speaking of his experiences at the Center for Advanced Study in the Behavioral Sciences, he says: "Particularly, I was struck by the number and extent of overt disagreements between social scientists about the nature of legitimate scientific problems and methods. Both history and acquaintance made me doubt that practitioners of the natural sciences possess firmer or more permanent answers to such questions than their colleagues in social science."

3. Thus, neither of the two most comprehensive British works on the sociology of scientific knowledge (Barnes, 1974; Bloor, 1976) quotes Merton's ([1945] 1973b) basic paper on the subject. At the same time, both authors are deeply influenced by Mary Douglas's neo-Durkheimism, which is not widely accepted among sociologists.

4. This interpretation of scientific conflict evidently differs from those prevalent in current philosophical literature (Kuhn, 1970; Lakatos, 1970) and accepted by most sociologists. That literature tends to take the metaphysical reasons given for the conflicts at their face value and to regard them as the basis of the conflict. According to the present interpretation, differences in problem choice and approach are often stated in metaphysical terms to make the differences *appear* reconcilable and thus to justify a conflict that may be useful as a weapon in the fight for resources. Evidence in support of this interpretation is provided by cases in which innovations giving rise to metaphysical conflict in one place are accommodated without a trace of such conflict at another. Thus psychoanalysis was assimilated into American academic psychology without any of the conflicts raised in continental Europe; another

example is the difference between the early history of bacteriology in Europe compared to its history in the United States and Japan.

References

Allen, T. J. "Communication Channels in the Transfer of Technology." *Industrial Management Review*, 1966, *8*, 87–98.

Allen, T. J., and Cohen, S. I. "Information Flow in Research and Development Laboratories." *Administrative Science Quarterly*, 1969, *14*, 12–19.

Barber, B. *Science and the Social Order*. New York: Free Press, 1952.

Barber, B. "Review of T. S. Kuhn. 'The Structure of Scientific Revolutions.'" *American Sociological Review*, 1963, *28*, 298–299.

Barnes, B. *Scientific Knowledge and Sociological Theory*. London: Routledge & Kegan Paul, 1974.

Barnes, B., and Dolby, R. G. A. "The Scientific Ethos: A Deviant Viewpoint." *European Journal of Sociology*, 1970, *2*, 3–25.

Ben-David, J. "Roles and Innovations in Medicine." *American Journal of Sociology*, 1960, *65*, 557–568.

Ben-David, J. *The Scientist's Role in Society: A Comparative Study*. Englewood Cliffs, N.J.: Prentice-Hall, 1971.

Ben-David, J. "Organization, Social Control, and Cognitive Change in Science." In J. Ben-David and T. N. Clark (Eds.), *Culture and Its Creators*. Chicago: University of Chicago Press, 1977.

Ben-David, J., and Collins, R. "Social Factors in the Origins of a New Science: The Case of Psychology." *American Sociological Review*, 1966, *31*, 451–465.

Bernal, J. D. *The Social Function of Science*. London: Routledge & Kegan Paul, 1939.

Bloor, D. *Knowledge and Social Imagery*. London: Routledge & Kegan Paul, 1976.

Blume, S. S., and Sinclair, R. "Chemists in British Universities: A Study of the Reward System in Science." *American Sociological Review*, 1973, *38*, 126–138.

Breiger, R. L. "Career Attributes and Network Structure: A Block Model Study of a Biomedical Research Speciality." *American Sociological Review*, 1976, *41*, 117–135.

Chubin, D. "The Conceptualization of Scientific Specialties." *Sociological Quarterly*, 1976, *17*, 448–476.

Clark, T. "Institutionalization of Innovations in Higher Education: Four Models." *Administrative Science Quarterly,* 1968, *13,* 1–25.

Clark, T. *Prophets and Patrons: The French University and the Emergence of the Social Sciences.* Cambridge, Mass.: Harvard University Press, 1973.

Cole, J. R., and Cole, S. *Social Stratification in Science.* Chicago: University of Chicago Press, 1973.

Cole, J. R., and Zuckerman, H. "The Emergence of a Scientific Specialty: The Self-Exemplifying Case of the Sociology of Science." In L. A. Coser (Ed.), *The Idea of Social Structure.* New York: Harcourt Brace Jovanovich, 1975.

Crane, D. "Social Structure in a Group of Scientists: A Test of the 'Invisible College' Hypothesis." *American Sociological Review,* 1969, *36,* 335–352.

Crane, D. *Invisible Colleges: Diffusion of Knowledge in Scientific Communities.* Chicago: University of Chicago Press, 1972.

Crawford, S. "Informal Communication Among Scientists in Sleep Research." *Journal of the American Society for Information Science,* 1971, *22,* 301–310.

Davis, K. *Human Society.* New York: Macmillan, 1949.

Dolby, R. G. A. "The Sociology of Knowledge in Natural Science." *Science Studies,* 1971, *1,* 3–21.

Fisher, C. S. "The Death of a Mathematical Theory: A Study in the Sociology of Knowledge." *Archives for History of Exact Sciences,* 1966, *3,* 137–159.

Gaston, J. *Originality and Competition in Science.* Chicago: University of Chicago Press, 1973.

Gaston, J. *The Reward System in British and American Science.* New York: Wiley-Interscience, 1978.

Glaser, B. G. *Organizational Scientists.* Indianapolis: Bobbs-Merrill, 1964.

Gordon, G., and Marquis, S. "Freedom, Visibility of Consequences and Scientific Innovation." *American Journal of Sociology,* 1966, *72,* 95–202.

Griffith, B. C., and Mullins, N. C. "Coherent Social Groups in Scientific Change." *Science,* 1972, 177, 959–964.

Griffith, B. C., and others. "The Structure of Scientific Literatures. II: Toward a Macro- and Microstructure for Science." *Science Studies,* 1974, *4,* 339–365.

Hagstrom, W. O. *The Scientific Community,* New York: Basic Books, 1965.

Holton, G. "Models for Understanding the Growth and Excellence of Scientific Research." In S. R. Graubard and G. Holton (Eds.), *Excellence and Leadership in a Democracy.* New York: Columbia University Press, 1962.

Kaplan, N. "Some Organizational Factors Affecting Creativity." *IRE Transactions on Engineering Management,* 1960, *EM-7,* 24–30.

Kaplan, N. "Organization: Will It Choke or Promote the Growth of Science?" In K. Hill (Ed.), *The Management of Scientists.* Boston: Beacon Press, 1964.

King, M. D. "Reason, Tradition and the Progressiveness of Science." *History and Theory,* 1971, *10,* 3–32.

Kornhauser, W. (with the assistance of W. O. Hagstrom). *Scientists in Industry.* Berkeley: University of California Press, 1962.

Krantz, D. "Do You Know What Your Neighbors Are Doing? A Study of Scientific Communication in Europe." *International Journal of Psychology,* 1970, *5,* 221–226.

Krantz, D. "The Separate Worlds of Operant and Non-operant Psychology." *Journal of Applied Behavior Analysis,* 1971, *4,* 61–70.

Kuhn, T. S. *The Structure of Scientific Revolutions.* Chicago: University of Chicago Press, 1962.

Lakatos, I. "Falsification and the Methodology of Scientific Research Programmes." In I. Lakatos and A. Musgrave (Eds.), *Criticism and the Growth of Knowledge,* Cambridge, England: Cambridge University Press, 1970.

Marcson, S. *The Scientist in American Industry.* New York: Harper & Row, 1960.

Martins, H. "The Kuhnian 'Revolution' and Its Implications for Sociology." In A. H. Nossiter, T. Hanson, and S. Rokkan (Eds.), *Imagination and Precision in Political Analysis.* London: Faber, 1971.

Merton, R. K. "Priorities in Scientific Discovery: A Chapter in the Sociology of Science." *American Sociological Review,* 1957, *22,* 635–659.

Merton, R. K. *Science, Technology and Society in Seventeenth-Century England* [1938a]. New York: Harper & Row, 1970.

Merton, R. K. "The Normative Structure of Science" [1942]. In

The Sociology of Science. Chicago: University of Chicago Press, 1973a.

Merton, R. K. "Paradigm for the Sociology of Knowledge" [1945]. In *The Sociology of Science.* Chicago: University of Chicago Press, 1973b.

Merton, R. K. "Science and the Social Order" [1938b]. In *The Sociology of Science.* Chicago: University of Chicago Press, 1973c.

Merton, R. K., and Gaston, J. (Eds.). *The Sociology of Science in Europe.* Carbondale: Southern Illinois University Press, 1977.

Mulkay, M. J. "Some Aspects of Growth in the Natural Sciences." *Social Research,* 1969, *36,* 22–52.

Mulkay, M. J. "The Sociology of Science in Britain." In R. K. Merton and J. Gaston (Eds.), *The Sociology of Science in Europe.* Carbondale: Southern Illinois University Press, 1977a.

Mulkay, M. J. "Sociology of the Scientific Research Community." In I. Spiegel-Rösing and D. de S. Price (Eds.), *Science, Technology and Society.* Beverly Hills, Calif.: Sage, 1977b.

Mullins, N. C. "The Development of a Scientific Specialty: The Phage Group and the Origin of Molecular Biology." *Minerva,* 1972, *10,* 51–82.

Mullins, N. C. *Theories and Theory Groups in Contemporary American Sociology.* New York: Harper & Row, 1973.

Mullins, N. C., and others. "The Group Structure of Co-Citation Clusters: A Comparative Study." *American Sociological Review,* 1977, *42,* 552–562.

Needham, J. (Ed.). *Science at the Crossroads: Papers Presented to the International Congress of the History of Science and Technology.* London: Cass, 1931.

Pelz, D., and Andrews, F., *Scientists in Organizations.* New York: Wiley, 1966.

Polanyi, M. *The Logic of Liberty: Reflexions and Rejoinders.* London: Routledge & Kegan Paul, 1951.

Price, D. de S. *Science Since Babylon.* New Haven, Conn.: Yale University Press, 1961.

Price, D. de S. *Little Science, Big Science.* New York: Columbia University Press, 1963.

Shepard, H. A. "Nine Dilemmas in Industrial Research." *Administrative Science Quarterly,* 1956, *1,* 295–309.

Shils, E. "Tradition, Ecology and Institution in the History of Sociology." *Daedalus,* 1970, *99,* 760–825.

Small, H. G., and Griffith, B. C. "The Structure of Scientific Literatures. I: Identifying and Graphing Specialties." *Science Studies,* 1974, *4,* 17–40.

Stehr, N. "Zur Soziologie der Wissenschaftssoziologie" ["On the Sociology of the Sociology of Science]. In N. Stehr and R. König (Eds.), *Wissenschaftssoziologie* [*Sociology of Science*]. Opladen, West Germany: Westdeutscher Verlag, 1975.

Storer, N. *The Social System of Science.* New York: Holt, Rinehart and Winston, 1966.

Whitley, R. D. "Communication Nets in Science: Status and Citation Patterns in Animal Physiology." *Sociological Review,* 1969, *17,* 219–234.

Zloczower, A. *Career Opportunities and the Growth of Scientific Discovery in 19th Century Germany, with Special Reference to Physiology.* Jerusalem: Hebrew University, 1960.

Index